D0938778

MEMORY STORAGE PATTERNS IN PARALLEL PROCESSING

THE KLUWER INTERNATIONAL SERIES IN ENGINEERING AND COMPUTER SCIENCE

PARALLEL PROCESSING AND FIFTH GENERATION COMPUTING

Consulting Editor

Doug DeGroot

Other books in the series:

PARALLEL EXECUTION OF LOGIC PROGRAMS
John S. Conery ISBN 0-89838-194-0

PARALLEL COMPUTATION AND COMPUTERS FOR ARTIFICIAL INTELLIGENCE
Janusz S. Kowalik ISBN 0-89838-227-0

SUPERCOMPUTER ARCHITECTURE
Paul B. Schneck ISBN 0-89838-234-4

ASSIGNMENT PROBLEMS IN PARALLEL AND DISTRIBUTED COMPUTING
Shahid H. Bokhari ISBN 0-89838-240-8

MEMORY STORAGE PATTERNS IN PARALLEL PROCESSING

by

Mary E. Mace
IBM Corporation

KLUWER ACADEMIC PUBLISHERS
Boston/Dordrecht/Lancaster

Distributors for North America:
Kluwer Academic Publishers
101 Philip Drive
Assinippi Park
Norwell, Massachusetts 02061 USA

Distributors for the UK and Ireland:
Kluwer Academic Publishers
MTP Press Limited
Falcon House, Queen Square
Lancaster LA1 1RN, UNITED KINGDOM

Distributors for all other countries:
Kluwer Academic Publishers Group
Distribution Centre
Post Office Box 322
3300 AH Dordrecht, THE NETHERLANDS

Library of Congress Cataloging-in-Publication Data

Mace, Mary E.
Memory storage patterns in parallel processing.

(Kluwer international series in engineering and computer science ; SECS 30)
Bibliography: p.
Includes index.
1. Parallel processing (Electronic computers)
2. Data structures (Computer science) I. Title.
II. Series.
QA76.5.M187546 1987 004'.35 87-17052
ISBN 0-89838-239-4

Printed in the United States of America

Contents

Preface

This project had its beginnings in the Fall of 1980. At that time Robert Wagner suggested that I investigate compiler optimization of data organization, suitable for use in a parallel or vector machine environment.

We developed a scheme in which the compiler, having knowledge of the machine's access patterns, does a global analysis of a program's operations, and automatically determines optimum organization for the data.

For example, for certain architectures and certain operations, large improvements in performance can be attained by storing a matrix in row major order. However a subsequent operation may require the matrix in column major order. A determination must be made whether or not it is the best solution globally to store the matrix in row order, column order, or even have two copies of it, each organized differently. We have developed two algorithms for making this determination.

The technique shows promise in a vector machine environment, particularly if memory interleaving is used. Supercomputers such as the Cray, the CDC Cyber 205, the IBM 3090, as well as superminis such as the Convex are possible environments for implementation.

Although we provide theoretical foundations, this monograph has been written with the implementer in mind. Our two algorithms are presented in chapters 2 and 3, and detailed examples in which our techniques are applied to entire program segments (not just individual operations) are provided in chapters 2,4, and 6. We hope that the reader will be able to use these techniques to improve performance in his or her particular environment.

Acknowledgements

This book would not have been possible without the assistance of several people.

Robert Wagner suggested the topic of this book as a Ph.D dissertation. As my adviser he was also responsible for many of the ideas presented here. Several years later he was kind enough to review a draft of the book.

Frank Starmer generously allowed me the use of his lab and computing facilities for the seven months it took to complete the project.

Kevin Bowyer reviewed a draft of the book on very short notice and made many helpful suggestions.

Anselmo Lastra provided invaluable assistance in the production and typesetting of the book.

Acknowledgments

MEMORY STORAGE PATTERNS
IN
PARALLEL PROCESSING

Chapter 1
INTRODUCTION

In this monograph the problem of selection of memory storage patterns or shapes is discussed and techniques for solution are presented. Memory storage patterns are important for efficiency of operation in a wide range of computing environments. The environment we will be discussing here is a parallel programming environment. Careful choice of memory storage patterns is particularly important in the efficient utilization of interleaved memory banks.[16, 18, 8, 15, 19]

Initially, a few concepts necessary to the discussion will be introduced:

- what are memory storage patterns?

- program graphs, and how they are used in the context of this discussion

- cost functions: the problem will be defined in terms of a cost function that is minimized.

- shared nodes in a program graph: we shall see that these nodes give the problem additional complexity that makes it more interesting mathematically.

In the second and third chapters two solution algorithms are presented. The algorithm in the second chapter is very efficient but restricted in the types of graphs it can handle. The algorithm in the third chapter, while less efficient, applies to a large class of program graphs. The relevant class of graphs is defined in the appendix. Also in the appendix can be found a proof that the algorithm handles the entire class of graphs.

In the fourth chapter the technique in chapter 3 is illustrated by an example, with a short code segment and hypothetical cost functions. The sixth chapter gives a more detailed example using the architecture of the CDC Cyber 205 and Jacobi iteration.

In the fifth chapter some of the mathematical complexity of the problem is illustrated. Several variations of the problem are reduced to known NP-complete problems.

What are shapes

The concept of memory storage patterns for data structures is quite general, so for this discussion we will restrict the data structures in question to one and two dimensional arrays (matrices). Also, for ease of discussion we will use the term *shapes* to refer to memory storage patterns.

As Figures 1 and 2 illustrate, a shape for a matrix A is an arrangement of the elements of A in memory.

Row order (Figure 1-1) is an example of a shape.

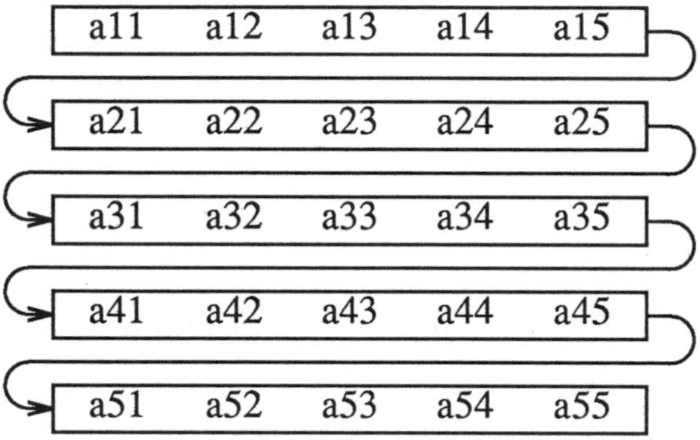

a11 a12 a13 a14 a15
a21 a22 a23 a24 a25
a31 a32 a33 a34 a35
a41 a42 a43 a44 a45
a51 a52 a53 a54 a55

Figure 1-1 Matrix A stored in row order

Column order (Figure 1-2) is a shape.

a11 a21 a31 a41 a51
a12 a22 a32 a42 a52
a13 a23 a33 a43 a53
a14 a24 a34 a44 a54
a15 a25 a35 a45 a55

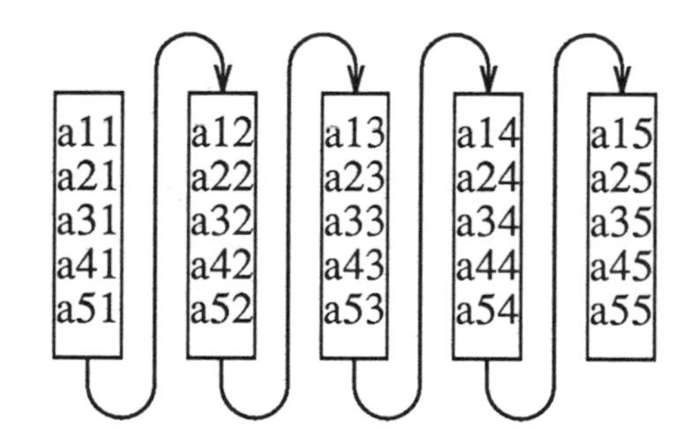

Figure 1-2 Matrix A stored in column order

Suppose the assumption is made that sequential (stride 1) access of the elements of A is more efficient than other types of access. By stride 1 we mean that we access element 0, followed by element 1,2,3. Stride n means accessing element n, followed by element 2n, 3n, and so on. For machines in which memory interleaving is used, the most efficient stride depends on how the addresses are organized among the interleaved memories. However for this example we assume stride 1.

For the matrix operation A × B, where × represents the inner product of A and B:

A in row order

B in column order

are efficient shapes for the × operation. Row order and column order are only two of many possible shapes. Some other examples of shapes are: skewed storage,[8, 18] storing only the diagonal of a diagonal matrix, and even sparse matrix representations.

Program Graphs

We now introduce the concept of program graphs. We assume the reader is familiar with the representation of a program or code segment as a set of nodes and arcs. However in this discussion the nodes and arcs are associated with the matrices, matrix operations, and shapes in a very precise way.

As Figure 1-3 illustrates, nodes represent operations: leaf nodes represent input operations, and internal nodes represent arithmetic operations such as + and ×. Arcs in the graph represent data (matrices).

A shape will be assigned to each arc in the program graph. For example, in the program graph in Figure 1-3 the following assignment can be made:

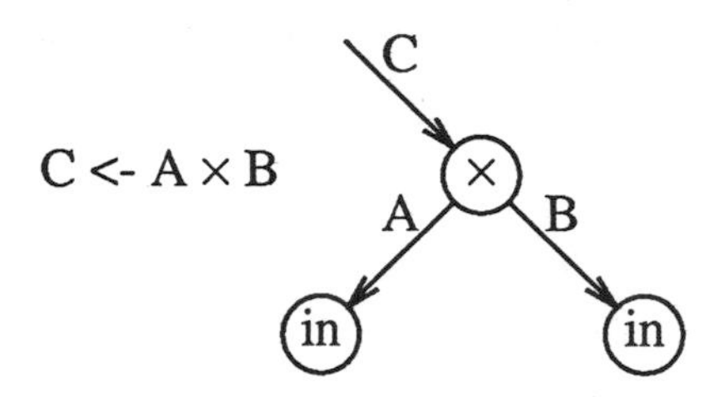

Figure 1-3 A program segment and its graph

- Assign row order to A
- Assign column order to B
- Assign row order to C

Cost functions

We would like to find shapes that make the matrix operations that we associate with nodes, more efficient. The approach we will take in order to do this is to associate a cost function with each node. This cost function is intended to represent efficiency (in terms of units of time) of the operation the node represents. The cost function will depend on two variables:

- Which operation the node represents: this is intended to represent the fact that some operations are much more efficient than others, and

- Which shapes are assigned to each of the arcs that touch that node.

Arcs touching the node correspond to the operands and result of the operation, so the cost of accessing these operands is a component of the cost of the operation.

The objective is to find a shape assignment for each arc in the program graph that minimizes the sum of the cost functions over all nodes in the graph.

Figure 1-4 illustrates the way we shall represent the cost function T_{op} for the operation $C = A\ op\ B$.

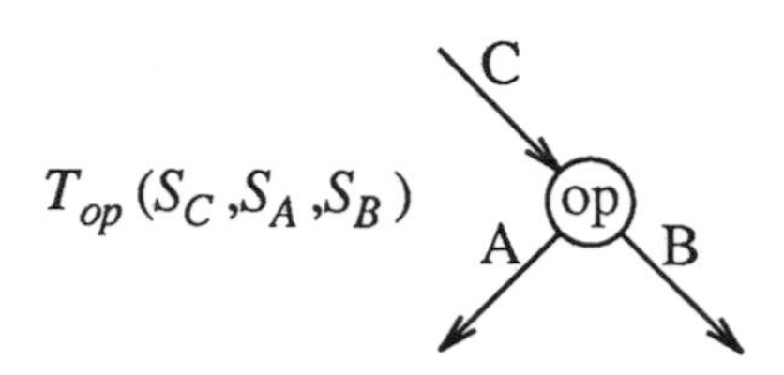

Figure 1-4 Cost function for a node

The node labeled "op" represents some operation with input operands A and B and result C. The variables S_C, S_A, and S_B represent the shapes which are assigned to arcs labeled A, B, and C.

We would like to choose S_C, S_A, and S_B so as to minimize T_{op}.[22,9,6]

Shared nodes

As the program graph in Figure 1-5 shows, an operand can be used in more than one operation.

Suppose one operation is most efficient with A in one shape, but the other operation is most efficient with A in a different

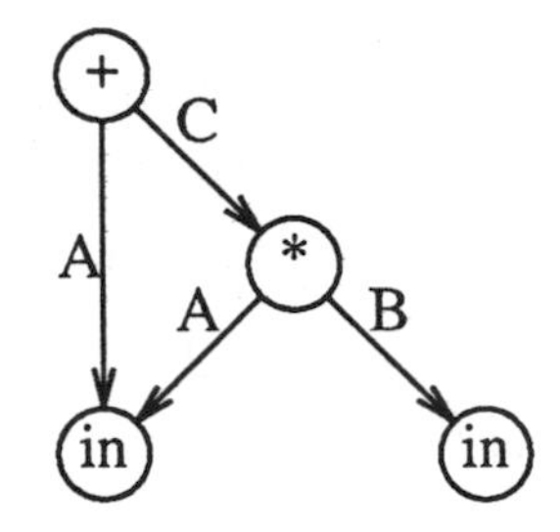

Figure 1-5 A shared node

shape.

We have several options:

Choose a shape for A that makes one operation efficient. The other operation may now be very inefficient.

Choose a compromise shape. Neither operation is as efficient as it might be, but neither is as inefficient as it might be.

Provide two copies of A by transposing. We use the word transpose here to mean transforming a matrix from one shape to another. A copy of A is provided in the shape that the one operation prefers. Another copy of A is provided in the shape the other operation prefers. However, an extra cost is incurred for the transpose operation. Note that all costs in this discussion refer to units of time. The reader can see clearly that the extra copy of A incurs a space penalty as well. The cost of space usage is not considered in this discussion. The problem of shared nodes has been studied extensively in the context of code generation for sequential machines.[5,21]

Objective of study

Now that several of the important concepts have been introduced, we can define the objective of this study more precisely.

Objective: find an assignment of shapes to all arcs in a graph which minimizes the total cost (the total cost refers to the sum of the costs of the individual nodes).

The general technique used to solve this problem is dynamic programming, involving graph reductions.

The techniques illustrated in the next several chapters can be applied to restricted classes of directed, acyclic graphs. Chapter two is limited to the case with no shared nodes in the graph. The class of graphs discussed in chapters three and four are called the collapsible graphs.[21, 17]

The collapsible graphs are defined precisely in Appendix A. In Appendix D we include a proof that the technique presented in chapter 3 can be applied to all collapsible graphs.

Background

While automatic selection of shapes by the compiler has not been applied, ad hoc shape selection by knowledgeable users has been done. The technique has been used successfully on vector processing architectures such as the CDC Cyber 205[15] and the IBM 3090.

The idea is to prearrange a large matrix in memory in the pattern that allows the computer to make optimal access for the particular problem being solved. The memory structure of the Cray machines, which leads to efficient access patterns of stride n, is also amenable to shape optimization by the user.

Selection of shapes has not yet been applied to multiprocessing architectures such as that of the Illinois Cedar project or the IBM RP3[20] although studies have been proposed for both machines.

Kascic[15] investigated the effect of optimal choice of matrix shapes for several numerical analysis applications on the CDC Cyber 205.

His study breaks down each algorithm into steps, and at each step looks at the effect of memory access of the relevant matrix or vector on the efficiency of that step. The objective was to choose the memory access pattern for each step which gave the greatest speedup for the program.

Applications such as Gaussian elimination, Jacobi iteration, Gauss-Seidel iteration and SOR iteration are demonstrably made more efficient in a vector computer environment by carefully thought out optimization of vector shapes.

Kascic's study is quite similar to the work in this monograph in its objective. The difference is that this monograph attempts to find an automatic procedure which could accomplish the same thing as Kascic's hand optimization.

Leuze[19] also studies memory access patterns in vector machines in relation to making program operations more efficient. He actually invents some new memory access patterns which are more tailored to the particular operations. His work is also similar to the work presented in this monograph in that the objective is to find the memory storage patterns which are the most efficient for the particular operations.

The difference is that the focus is on individual operations rather than considering all the operations using matrices and vectors and their relationship to each other.

The algorithms used for automatic shape selection resemble certain dynamic programming algorithms for optimal code generation.

Optimal Code Generation has been studied for several different classes of program graphs and different abstract machine models.[7,21,5,1,2,3] A solution to the optimal code generation problem for expression trees has been found for a register machine model[1] and for a stack machine model.[7] Problems with this solution, such as the use of register pairs, have also been found.[2] The optimal code generation problem has been shown to be NP-complete when common subexpressions are allowed in the graph.[5]

The concept of collapsible graphs comes from Prabhala and Sethi.[21] The authors were able to find an algorithm for optimal code generation[5] in polynomial time for the class of collapsible graphs. Since common subexpression elimination gives rise to program graphs with shared nodes, code generation for the entire class of directed acyclic graphs with shared nodes is NP-hard. This gave us a clue that the apparently very complex shapes problem might also be able to be solved efficiently for this restricted but very large class of graphs.

Nonserial dynamic programming[6] is a modeling and solution technique that can be applied directly to solve the shapes problem. No upper bound on time has been found for this process, however. The technique of graph reductions combined with cost function minimization used in the collapsible graph algorithm resembles to some extent the technique of nonserial dynamic programming.

The existence of the field of compiler optimization for

parallel environments is due to Kuck and the other researchers at the University of Illinois. Many types of optimizations are discussed in[16] (not shapes however). The best known and widely used results of this research is data dependence analysis of DO loops, which determines whether the statements of a DO loop may be executed in parallel. We consider this work to be the foundation of the art of compiling for parallelism.

Chapter 2

SOLUTION FOR GRAPHS WITHOUT SHARED NODES

A tree is a directed acyclic graph with no shared nodes. In a tree every node has exactly one immediate ancestor. If the graph representing a program is a tree, then each result is used in exactly one operation after it is computed. Below is a diagram of a tree (left) and a graph which is not a tree (right).

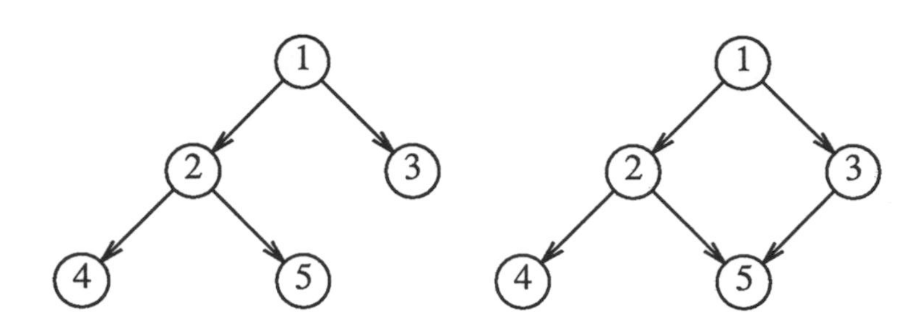

Figure 2-1 Directed acyclic graph which is a tree (left) and D.A.G. which is not a tree due to a shared node (right).

The problem of determining the best shapes for shared node arcs was discussed briefly in the previous chapter. It should come as no surprise that an efficient algorithm for the solution of the shapes problem was discovered for the class of trees, since they contain no shared nodes. As the reader will see, the tree algorithm is a dynamic programming algorithm which takes quadratic time. The tree algorithm resembles in its general style the code generation algorithm in Aho and Johnson.[1]

Concepts necessary to the tree algorithm

Before presenting the tree algorithm, two concepts fundamental to it will be explained. One is the concept of different methods, with different costs which can be chosen to do a particular operation.

The other concept that needs to be described is the operation of transposing from one shape to another.

Methods

We will assume that a particular operation, represented by a node, has one left operand, one right operand, and one result. We will also assume that there are one or more *methods* from which the computer can choose to do the operation. For example, two methods which could be chosen for matrix multiplication are inner product and outer product. The following strict assumptions will be made about these methods.

- Each method can have a different cost. The cost can be thought of as the units of time necessary to do the operation. (see chapter 6 for detailed examples)

- Each method requires that its left operand be in a certain shape, and its right operand be in a certain shape.

- Each method leaves its result in a certain shape.

These specifications are assumed to be known, and stored in a large table. This table describes the dependency of the cost of the operation on the shapes of its operands and result. If the left and right operands for a node are in the shapes required by a particular method then that method can be used. The cost of the computation is then the cost of that method.

For example, suppose M_1 and M_2 are methods, each of which does a particular operation, such as matrix multiplication. M_1 could require that its left operand be in shape S_1, its right operand be in shape S_2, its result be in shape S_3, and have a cost of 10. M_2 could require both left and right operands to be in shape S_1, and also leave its result in shape S_1, and have a cost of 5.

This data would be represented in a table as a quintuple, whose components are <method, result shape, left operand shape,

right operand shape, cost> and would appear as the quintuples $<M_1, S_3, S_1, S_2, 10>$ and $<M_2, S_1, S_1, S_1, 5>$.

Transposing from one shape to another

Given the availability of two shapes S_1 and S_2, and two methods M_1 and M_2, suppose one wanted to find the least expensive way to compute a result and have it be in shape S_1. Suppose method M_1 leaves its result in shape S_1 and its cost is 10. Suppose method M_2 leaves its result in shape S_2 and has a cost of 2. Finally, suppose the cost of changing the result shape from S_2 to S_1 is 1. The least expensive way to get the result into shape S_1 is obviously to use M_2 to compute the result into shape S_2 and then change it into shape S_1. The operation of simply changing a result's shape we term *transposing* the result (see Table 2-1).

In general, in order to find the minimum cost way to get a result into a particular shape S_1, it is necessary to examine the cost of not only computing it into S_1 but also computing it into every other shape and transposing into S_1. Transposing a matrix from one shape to another is also assumed to have a cost in terms of time units.

Figure 2-2 illustrates the process by which either of two methods computes a result, followed by a transposition into each of three possible shapes. Each path from a method to a particular target shape represents a feasible combination of method and transposition yielding that shape. The algorithm must choose the particular path which minimizes the cost.

input shape	output shape	cost
S_1	S_1	0
S_1	S_2	1
S_1	S_3	1
S_2	S_1	1
S_2	S_2	0
S_2	S_3	1
S_3	S_1	1
S_3	S_2	1
S_3	S_3	0

Table 2-1 Costs for transposing three shapes

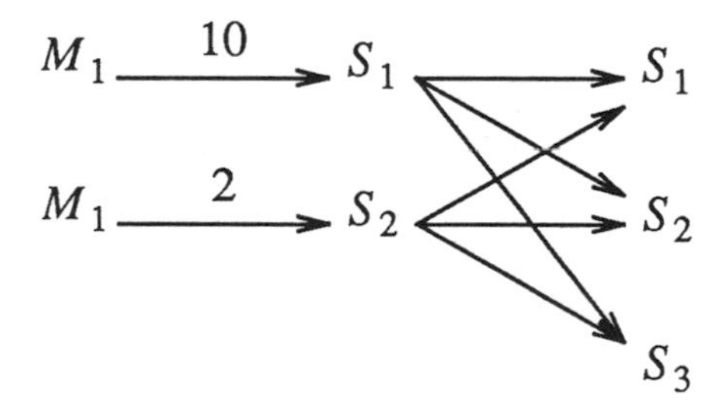

Figure 2-2 The two steps of the tree Algorithm, computing and transposing

The cost table

For every node in the graph, and every possible shape the result of that node can be in, we compute a cost, and store it in a table. The tree algorithm finds a minimum cost for computing node i, and for leaving it in every possible shape.

The cost for a node i is computed by summing:

1) The cost of the operation, using some method m.

2) The minimum cost of computing the left operand, and leaving its result in the shape required by method m.

3) The minimum cost of computing the right operand, and leaving its result in the shape required by method m.

This computation is done by scanning the set of possible methods for an operation, summing the three costs described above, and choosing the sum with the lowest cost. When the cost for a node i is computed as above, we must have items 2) and 3) already computed. The algorithm starts by initializing the leaf nodes of the graph with input costs, and then computing the cost for each internal node only after the costs of its two descendants are available.

The cost table stores the cost of computing a node into every possible result shape. Since the available methods for the operation may not produce all possible result shapes a transposition step may be necessary. Table 2-2 illustrates three nodes, each of which can have its result in one of three shapes. The costs in the table are hypothetical.

Node	Result Shape		
	S_1	S_2	S_3
1	2	10	3
2	5	1	9
3	8	6	7

Table 2-2 Table of costs for three nodes, with results in three shapes

Variable definitions for tree algorithm

Let each node of the tree be labeled with a unique number. Let i stand for a node which represents a computation. The incoming arc to node i represents the result of node i's computation, and is also labeled i. The shape assigned to arc i will be termed S_i. *Left* (i) and *right* (i) are the nodes which represent the left and right operands of node i. Let j be *left*(i) and k be *right* (i).

Let *transposecost* (S_i, S_j) be the cost of transposing a data item from shape S_i to shape S_j. It is assumed that *transposecost* (S_i, S_i) is zero.

Each node in the graph has a unique number. For each node i, an array of costs is computed. This array of costs for node i is indexed by the set of shapes assignable to the arc which represents the result of node i's computation.

For each shape of result i, the cost corresponding to that shape represents the minimum cost of computing node i into that shape.

If we assume for every arc in the graph that the same set of s

shapes can be assigned, the cost table is s by n: a two dimensional table for getting each node's result into each possible shape. (Both of the algorithms in this monograph generalize to the situation in which a different set of shapes is assignable to the distinct arcs.)

This table will be called $cost(n, s)$, where n represents the total number of nodes and s represents the total number of shapes assignable to any arc.

In the first step of the minimization procedure described below, each of the methods which can be applied to node i's operation are considered. For each method M_p the following three items are added:

- The cost of doing M_p, or $M_p.opcost$

- The cost of getting the left operand j into the shape which M_p requires

- The cost of getting the right operand k into the shape which M_p requires.

The shape required for the left operand of M_p is $M_p.leftshape$. If the cost of getting node j into result shape S_j is $cost(j, S_j)$, the cost of getting node j into $M_p.leftshape$ is $cost(j, M_p.leftshape)$.

The shape required for the right operand of M_p is $M_p.rightshape$. If the cost of getting k into shape S_k is $cost(k, S_k)$, the cost of getting operand k into $M_p.rightshape$ is $cost(k, M_p.rightshape)$.

In the second step of the procedure the possibility of achieving the target shape by computing a result into every other shape and transposing is considered. As described above, this method may yield a lower cost.

Tree algorithm

The following is the recursive algorithm for computing the cost table for each node of the graph.

If node i is a leaf, $cost(i, S_i)$ is 0.

If node i is an operator node, $cost(i, S_i)$ is computed as follows:

for each shape S_i which may be assigned to arc i,

$$
\begin{aligned}
temp(S_i) = \min\{&(M_1.opcost + \\
&\quad cost(j, M_1.leftshape) + \\
&\quad cost(k, M_1.rightshape)), \\
&(M_2.opcost + \\
&\quad cost(j, M_2.leftshape) + \\
&\quad cost(k, M_2.rightshape)), \\
&\cdot \\
&\cdot \\
&\cdot \\
&(M_m.opcost + \\
&\quad cost(j, M_m.leftshape) + \\
&\quad cost(k, M_m.rightshape))\}
\end{aligned}
$$

The total time for this step is m, the total number of methods, for each node. For each shape S_j that can be assigned to arc a:

$$
\begin{aligned}
cost(a, S_j) = \min\{&(temp(S_1) + transposecost(S_1, S_j)), \\
&(temp(S_2) + transposecost(S_2, S_j)), \\
&\quad\vdots \\
&(temp(S_s) + transposecost(S_s, S_j))\}
\end{aligned}
$$

The total time for this step is s^2, where s is the total number of shapes. The total time for the tree algorithm to solve the shapes problem on a tree with n nodes is proportional to $n \times (m + s^2)$, or $\mathbf{O}((m + s^2)\, n)$.

Tree algorithm example

In this example we will use the tree algorithm to find the shape assignment which gives the lowest cost for the graph in Figure 2-3. We will only have two available shapes, S_1 and S_2. The cost of transposing from S_1 to S_2 will be 2 for this example.

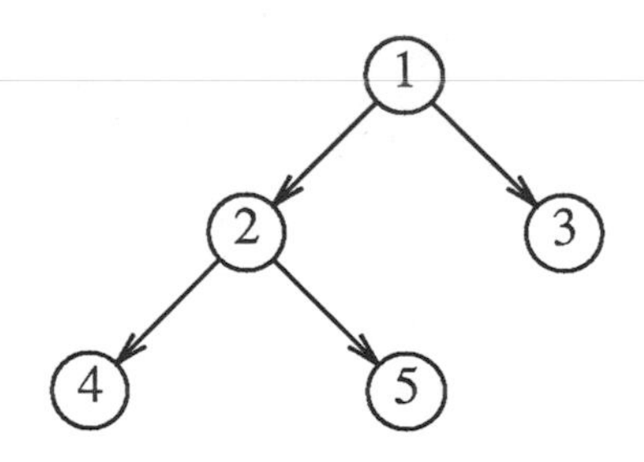

Figure 2-3 Tree algorithm example

We will compute these costs from the bottom up since the tree algorithm requires the costs of the left and right children of a node to compute that node's cost. The minimum cost for computing the result of each node into S_1 and S_2 will be stored in a table, illustrated in table 2-1. Nodes 3, 4, and 5 are considered to be leaf nodes. The cost of getting each of nodes 3, 4, and 5 into shapes S_1 and S_2 will be 0. The costs of getting nodes 4 and 5 into every available shape is necessary before the cost for node 2 can be computed. Similarly, the costs of getting nodes 2 and 3 into every shape is necessary before the cost for node 1 can be computed. Table 2-3 shows the table of costs for each node and each shape before the computation for nodes 1 and 2.

	S_1	S_2
1		
2		
3	0	0
4	0	0
5	0	0

Table 2-3 Cost table with costs for leaf nodes 3,4,and 5.

Node 2 can be computed by two methods, M_{21} and M_{22}.

M_{21} requires its left operand in S_1, its right operand in S_2, leaves its result in S_1 and has an opcost of 4. We will apply the formula given earlier.

$$Tcost = cost[leftop, M_{21}.leftopshape] + cost[rightop, M_{21}.rightopshape] + M_{21}.opcost.$$

$$Tcost = cost[4, S_1] + cost[5, S_1] + M_{21}.opcost$$

$$Tcost = 0 + 0 + 4$$

$$Tcost = 4.$$

The second part of the formula gives the cost of getting the result into every shape via transposing. This is shown by Figure 2-4.

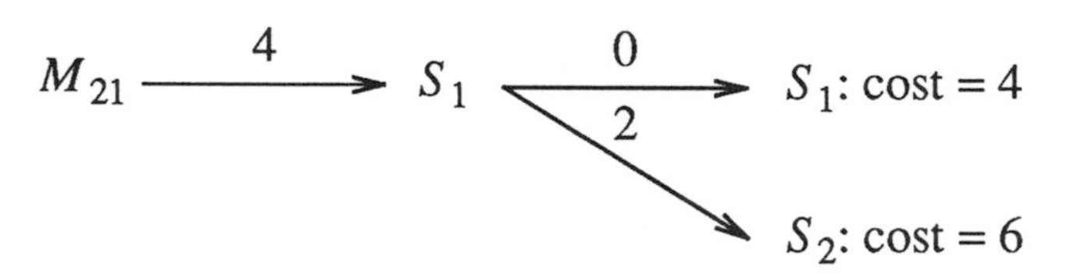

Figure 2-4 Cost of getting node 2 into S_1 *and* S_2 *using* M_{21}

M_{22} requires its left operand in S_2, its right operand in S_2, leaves its result in S_2 and has opcost = 3.

$$Tcost = cost[4, S_2] + cost[5, S_2] + M_{22}.opcost$$

$$Tcost = 0 + 0 + 3$$

$$Tcost = 3$$

Figure 2-5 shows the additional cost of transposing.

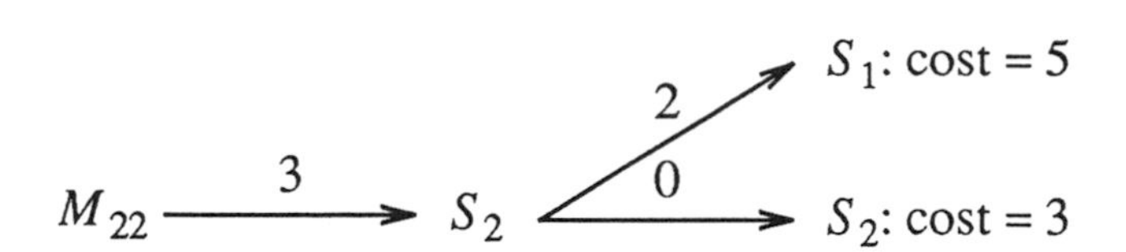

Figure 2-5 Cost of getting node 2 into S_1 and S_2 using M_{22}

Now we can update our cost table for node 2. The least cost for getting node 2 into S_1 is 4, using M_{21}. The least cost for getting node 2 into S_2 is 3, using M_{22}. The updated table follows.

	S_1	S_2
1		
2	4	3
3	0	0
4	0	0
5	0	0

Table 2-4 Cost table with costs for nodes 2, 3, 4, and 5.

Node 1 can be computed by two methods, M_{11} and M_{12}.

M_{11} requires its left operand in S_1, its right operand in S_2, leaves its result in S_1, and has opcost = 3.

$$Tcost = cost[2, S_1] + cost[3, S_2] + opcost$$

$$Tcost = 4 + 0 + 3$$

$$Tcost = 7$$

As above, we will illustrate the additional transpose cost in a figure.

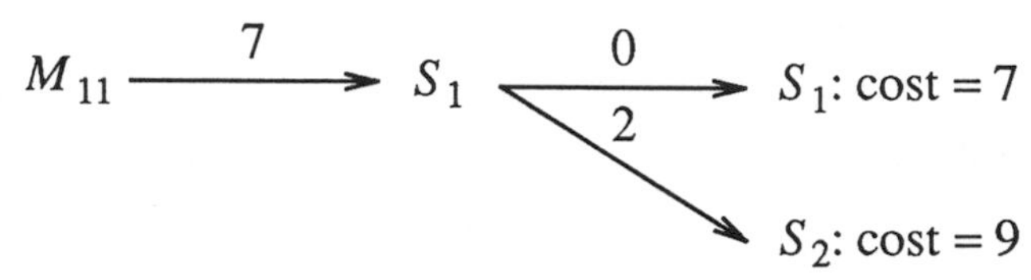

Figure 2-6 Cost of getting node 1 into S_1 and S_2 using M_{11}

M_{12} requires its left operand in S_2, its right operand in S_2, computes its result into S_2 and has opcost = 5.

$$Tcost = cost[2, S_2] + cost[3, S_2] + opcost$$

$$Tcost = 3 + 0 + 5$$

$$Tcost = 8$$

The additional cost of transposing is shown in Figure 2-7.

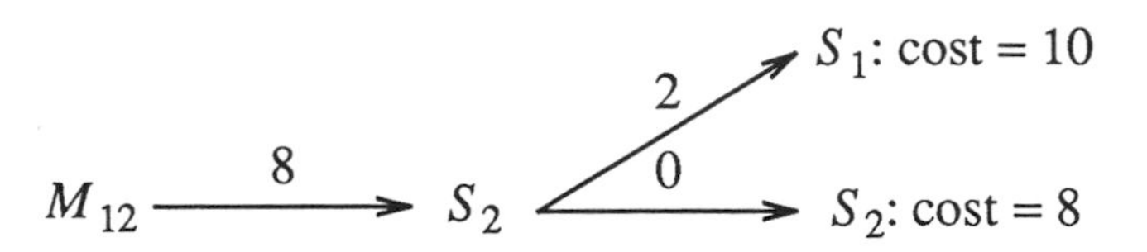

Figure 2-7 Cost of getting node 1 into S_1 and S_2 using M_{12}

Now we can update our cost table for node 1. The least cost for getting node 1 into S_1 is 7, using M_{11}. The least cost for getting node 1 into S_2 is 8, using M_{12}. The completed cost table is shown below.

	S_1	S_2
1	7	8
2	4	3
3	0	0
4	0	0
5	0	0

Table 2-5 Cost table with costs for all nodes

The reader might note that the cost table we have constructed shows only the cost for the nodes, not the method which gives the cost, nor the shapes the operands need to be in to use the method. This information can either be stored in a separate data structure as it is computed, or reconstructed from the cost table and another pass over the tree.

Chapter 3

SOLUTION FOR GRAPHS WITH SHARED NODES

In this section we describe a technique for finding an assignment of shapes to arcs which minimizes the total cost of a program graph. The technique depends on the graph having a certain structure which we call collapsible. A collapsible graph is a directed, acyclic graph which is series-parallel and includes treelike subgraphs.[17,21,12] A more precise definition of the collapsible graphs, which may be found in Appendix A, is not necessary to understand the collapsible graph algorithm.

In general, the solution algorithm involves a sequence of reductions on the graph, each reduction replacing two nodes with one and removing from the graph all arcs that connect the two nodes. The algorithm terminates when the graph has been reduced to a single node.

The cost function of the new single node is the sum of the cost functions of the two original nodes. The shape of the arc(s) which connected the two nodes is chosen such that the sum of the two cost functions is minimized.

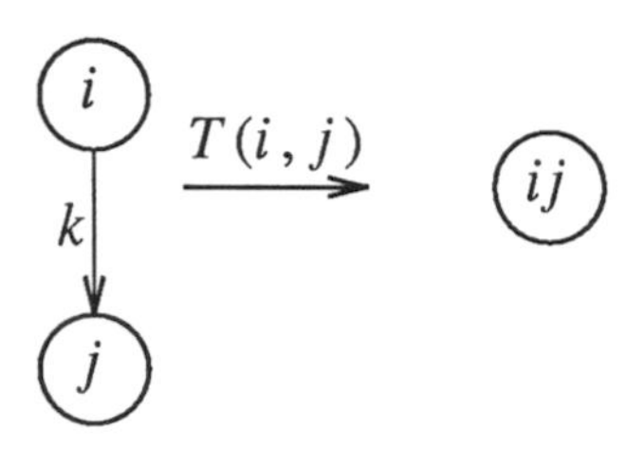

Figure 3-1 A graph transformation

In Figure 3-1, node i has cost function $T_i(S_k)$ which means that the cost of executing node i depends on the shape assigned to arc k, which is a variable. Node j has the cost function $T_j(S_k)$ which also depends on S_k. The cost for the new node ij is the sum of $T_i(S_k)$ and $T_j(S_k)$. A shape S_k for arc k is chosen such that the sum of the two cost functions T_i and T_j is minimized. When the graph has been reduced to a single node and there are no arcs remaining, an assignment of shapes will have been found that minimizes the total cost of the graph.

Upper bound on time required for algorithm

Below we give an algorithm which solves the shapes problem for collapsible graphs in time proportional to $(n-1) \times s^{d+1}$. Here, n is the number of nodes in the graph, s is the number of possible shapes, and d is the degree of the node with maximum degree. For fixed d this time bound is polynomial (see Appendix E for proof of time bound).

A restriction on the collapsible graphs

The following further restriction is placed on the class of collapsible graphs in order to derive the time bound on execution of this algorithm. The restriction is that each node must have indegree ≤ 1 or outdegree ≤ 1. This restriction will be referred to as *Restriction 1*.

Any collapsible graph G that violates Restriction 1 and that represents a program can be changed to a graph G' that satisfies it (see Appendix B for discussion).

Transformations A and B

The claim is that a sequence of graph transformations A and B (defined below) reduce to a single node any collapsible graph which meets Restriction 1. In the process of collapsing the graph, the minimum cost shape assignment is found. The definitions and figures below describe the circumstances under which transformations A and B can be applied, and the result of applying them, both in terms of graphs and cost functions.

In Figures 3-2 through 3-5 any arc which enters or leaves the subgraph before the transformation will enter or leave the node that is formed as a result of the transformation. The set of shapes assignable to that arc is one of the indices of the table created for the new node.

Let T_j be the cost function for a node j. The variables on which T_j depends are the shapes assigned to all arcs which touch node j. These arcs are called $j.1 \cdots j.t$. The variables representing shapes assignable to arcs $j.1 \cdots j.t$ are $S_{j.1} \cdots S_{j.t}$.

The notation $A(i, j)$ and $B(i, j)$ indicates that nodes i and j are being merged by A and B. In discussions where the particular nodes do not matter, the transformations will be called A and B.

Transformation $A(i, j)$ can be applied to a subgraph with nodes i and j such that $outdeg(j) = 0$, and node i is the only parent of node j, although node i may have other children besides node j. There are t $(t \geq 1)$ arcs from i to j.

$A(i, j)$ creates a subgraph consisting of a single node ij. $Indeg(ij)$ is the same as $indeg(i)$ in the original graph. $Outdeg(ij)$ is the same as $outdeg(i)$ in the original graph minus the t arcs that went to j.

The equation below combines the cost functions of node i, $T_i[a, S_{j.1} \cdots S_{j.t}, b1 \cdots bn]$ and node j, $T_j[S_{j.1} \cdots S_{j.t}]$. The sum finds the shapes for arcs $j.1 \cdots j.t$ that give a minimum cost relative to the shapes on arcs $a, b1 ... bn$. $S_{j.1}^1$ represents the assignment of shape 1 to arc $j.1$.

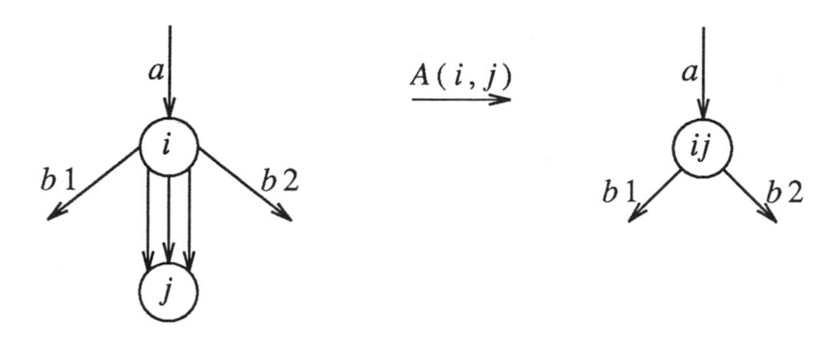

Figure 3-2 Transformation A

$$
\begin{aligned}
T_{ij}[a, b_1, \cdots b_n] = \min\{ \\
&(T_i[a, S^1_{j.1}, S^1_{j.2}, \cdots S^1_{j.t}, b_1, \cdots b_n] + T_j[S^1_{j.1}, S^1_{j.2}, \cdots S^1_{j.t}]), \\
&(T_i[a, S^2_{j.1}, S^1_{j.2}, \cdots S^1_{j.t}, b_1, \cdots b_n] + T_j[S^2_{j.1}, S^1_{j.2}, \cdots S^1_{j.t}]), \\
&\quad \vdots \\
&(T_i[a, S^h_{j.1}, S^1_{j.2}, \cdots S^1_{j.t}, b_1, \cdots b_n] + T_j[S^h_{j.1}, S^1_{j.2}, \cdots S^1_{j.t}]), \\
&(T_i[a, S^1_{j.1}, S^2_{j.2}, \cdots S^1_{j.t}, b_1, \cdots b_n] + T_j[S^1_{j.1}, S^2_{j.2}, \cdots S^1_{j.t}]), \\
&\quad \vdots \\
&(T_i[a, S^h_{j.1}, S^h_{j.2}, \cdots S^h_{j.t}, b_1, \cdots b_n] + T_j[S^h_{j.1}, S^h_{j.2}, \cdots S^1_{j.t}]) \}
\end{aligned}
$$

Transformation $B(i, j)$ can be applied to a subgraph consisting of two nodes i and j such that i is a parent of j and $t \geq 1$ arcs connect i with j. Slight variations in the subgraph account for 3 slightly different cases of transformation $B(i, j)$. The significant difference between transformation A and transformation B is that in transformation A node j is a leaf node, or node with no descendants. In the three cases of transformation B, node j is an internal node, or a node with one or more children.

Case 1: j is i's only child; i has indegree ≤ 1, outdegree $= t$; and j has indegree $= q \geq t$, outdegree ≤ 1. In case 1, i has only one child, but j may have other parents besides i.

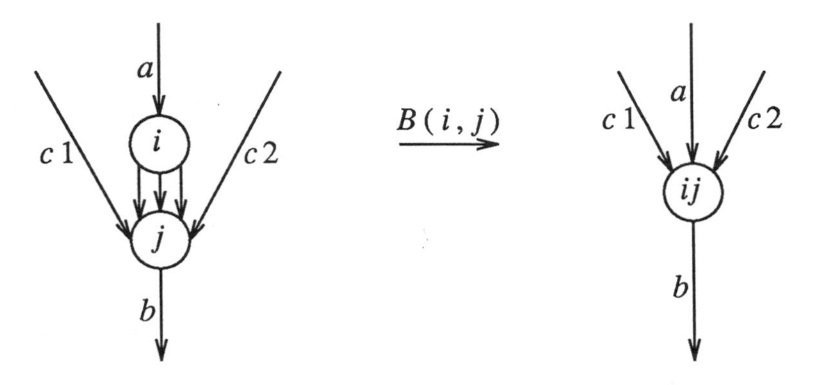

Figure 3-3 Transformation B, Case 1

$$T_{ij}[a, b, c_{1,} \cdots c_{q-t}] = \min \{$$

$$(T_i[a, S_{j.1}^1, S_{j.2}^1, \cdots S_{j.t}^1] + T_j[b, c_1, \cdots c_{q-t}, S_{j.1}^1, S_{j.2}^1, \cdots S_{j.t}^1$$

$$(T_i[a, S_{j.1}^2, S_{j.2}^1, \cdots S_{j.t}^1] + T_j[b, c_1, \cdots c_{q-t}, S_{j.1}^2, S_{j.2}^1, \cdots S_{j.t}^1$$

$$\vdots$$

$$(T_i[a, S_{j.1}^h, S_{j.2}^1, \cdots S_{j.t}^1] + T_j[b, c_1, \cdots c_{q-t}, S_{j.1}^h, S_{j.2}^1, \cdots S_{j.t}^1$$

$$(T_i[a, S_{j.1}^1, S_{j.2}^2, \cdots S_{j.t}^1] + T_j[b, c_1, \cdots c_{q-t}, S_{j.1}^1, S_{j.2}^2, \cdots S_{j.t}^1$$

$$\vdots$$

$$(T_i[a, S_{j.1}^h, S_{j.2}^h, \cdots S_{j.t}^h] + T_j[b, c_1, \cdots c_{q-t}, S_{j.1}^h, S_{j.2}^h, \cdots S_{j.t}^h$$

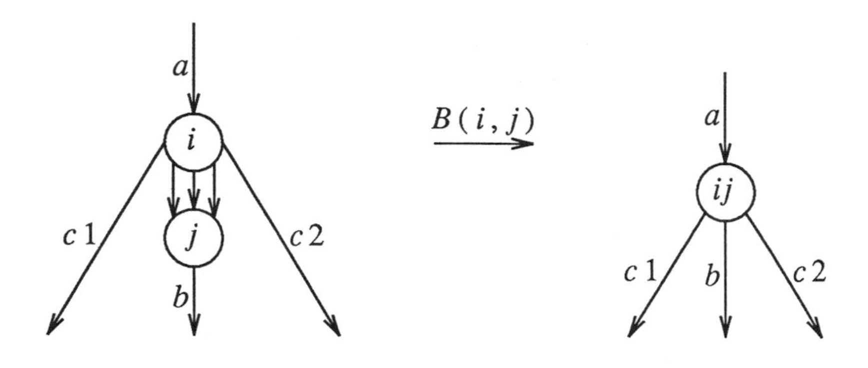

Figure 3-4 Transformation B, Case 2

Case 2: i is j's only parent; i has indegree ≤ 1, and outdegree $= q \geq t$; j has indegree $= t$, outdegree $= 1$. In case 2, j only has i as a parent, while i can have other children.

$$
\begin{aligned}
T_{ij}[a, b, c_{1,} \cdots c_{q-t}] = \min \{ \\
&(T_i[a, c_1 \cdots c_{q-t}, S_{j.1}^1, S_{j.2}^1, \cdots S_{j.t}^1] + T_j[b, S_{j.1}^1, S_{j.2}^1, \cdots S_{j.t}^1]), \\
&(T_i[a, c_1 \cdots c_{q-t}, S_{j.1}^2, S_{j.2}^1, \cdots S_{j.t}^1] + T_j[b, S_{j.1}^2, S_{j.2}^1, \cdots S_{j.t}^1]), \\
&\quad \vdots \\
&(T_i[a, c_1 \cdots c_{q-t}, S_{j.1}^h, S_{j.2}^1, \cdots S_{j.t}^1] + T_j[b, S_{j.1}^h, S_{j.2}^1, \cdots S_{j.t}^1]), \\
&(T_i[a, c_1 \cdots c_{q-t}, S_{j.1}^1, S_{j.2}^2, \cdots S_{j.t}^1] + T_j[b, S_{j.1}^1, S_{j.2}^2, \cdots S_{j.t}^1]), \\
&\quad \vdots \\
&(T_j[a, c_1 \cdots c_{q-t}, S_{j.1}^h, S_{j.2}^h, \cdots S_{j.t}^h] + T_j[b, S_{j.1}^h, S_{j.2}^h, \cdots S_{j.t}^h]) \}
\end{aligned}
$$

Case 3: $t = 1$; $indeg(i) \leq 1$, $outdeg(j) = 1$, i is j's only parent. The result is a subgraph consisting of a single node ij. The arc which joined i and j is deleted. The indegree of the new node is: $indeg(i) + indeg(j) - t$ and the outdegree is $outdeg(i) + outdeg(j) - t$. In case 3, there is only one arc connecting i and j.

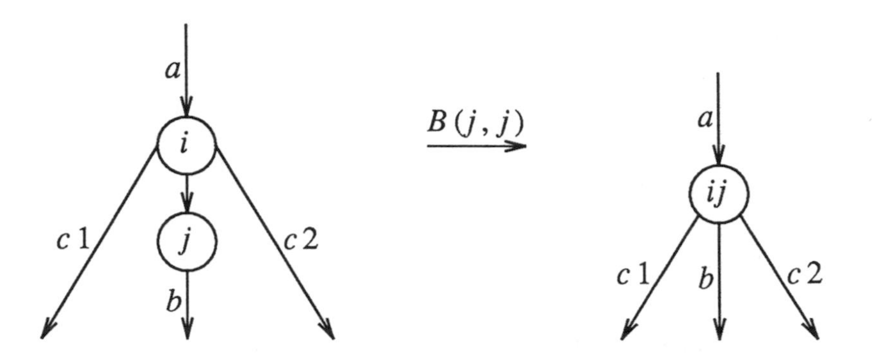

Figure 3-5 Transformation B, Case 3

$$\begin{aligned} T_{ij}[a, b, c_1 \cdots c_{q-1}] = \min \{ & (T_i[a, S_{j.1}^1, c_1 \cdots c_{q-1}] + T_j[b, S_{j.1}^1]), \\ & (T_i[a, S_{j.1}^2, c_1 \cdots c_{q-1}] + T_j[b, S_{j.1}^2]), \\ & \vdots \\ & (T_i[a, S_{j.1}^h, c_1 \cdots c_{q-1}] + T_j[b, S_{j.1}^h]) \} \end{aligned}$$

The following chapter will illustrate a sequence of transformations A and B being applied to a graph with shapes and cost functions, resulting in an assignment of shapes to the program data.

Chapter 4

ILLUSTRATION OF COLLAPSIBLE GRAPH ALGORITHM

In this section we will present an example of how the collapsible graph algorithm works. Our illustration uses a short, simple code segment involving matrix operands, however the program graph of this code segment contains a shared node. A memory model is provided, from which meaningful cost functions for the program operations are derived. These cost functions are illustrated as tables indexed by the shapes of operands and results. We then go through a process of three graph reductions, eventually reducing the graph to a single node, and in the process finding a shape assignment for each matrix operand.

Program segment

In the program segment in figure 4-1 all operands are M × M matrices and all operations are matrix operations.

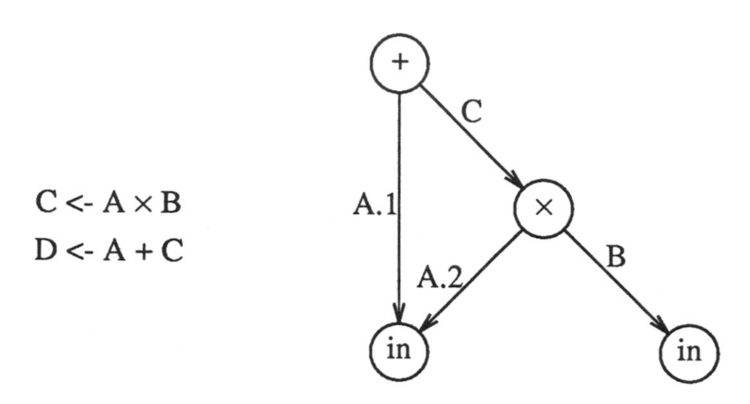

Figure 4-1 A program segment

Two points of interest should be noted. First, operations × and + have different shape requirements for their operands. This will be seen clearly in the cost tables in figures 4-6 and 4-9. Second, operand A is shared. The two arcs representing A are labeled A.1 and A.2. Each of these may be assigned a different shape. Finally, in order to avoid confusion the cost table for each operation will be labeled with the letter corresponding to the operand resulting from that operation. The cost table for the operation that reads operand B will be labeled T_B. The cost table for the operation that reads operand A will be labeled T_A. The cost table for the × operation, which computes operand C will be labeled T_C. The cost table for the + operation, which computes operand D will be labeled T_D. When two nodes are combined, the new node and new cost table are renamed by combining the letters corresponding to the result operands of the two nodes. In some cases the multi-letter labeling of some nodes and cost tables has been shortened arbitrarily. Despite this, it should be clear from context in each case which nodes are being combined.

Memory model

In the memory model shown in figure 4-2, the system has M interleaved memory modules. If each element of a vector less than or equal to M elements long, referred to as an M vector, is stored in a separate unit, conflict free access is possible.[8, 18]

0	1	2	3	4
a_{11}	a_{12}	a_{13}	a_{14}	a_{15}
a_{21}	a_{22}	a_{23}	a_{24}	a_{25}
a_{31}	a_{32}	a_{33}	a_{34}	a_{35}
a_{41}	a_{42}	a_{43}	a_{44}	a_{45}
a_{51}	a_{52}	a_{53}	a_{54}	a_{55}

Figure 4-2 A memory model

The five elements $\{a_{11}, a_{12}, a_{13}, a_{14}, a_{15}\}$ can be accessed without conflict since each is located in a separate memory module. The five elements $\{a_{11}, a_{21}, a_{31}, a_{41}, a_{51}\}$ are all located in memory bank zero, so accessing them would be more costly in terms of memory conflicts.

Derivation of cost functions

Given the memory model we can now derive cost functions. The cost of the × and + operations is based on the number of conflicts that occur while accessing the necessary M vectors. The difference in time required for the two operations themselves will not be considered in this discussion. All units are units of time.

1. Each conflict free access of an M vector has a cost of 10 units.

2. If accessing an M vector causes n memory conflicts, the cost of an operation is multiplied by n.

3. An extra cost of 10 units is charged for transposing an operand into another shape.

Figure 4-3 illustrates a conflict free access of the M vector $\{a_{11}, a_{12}, a_{13}, a_{14}, a_{15}\}$. Accessing this vector has a cost of 10 units from rule 1 above. Figure 4-4 illustrates an M vector stored in such a way that accessing it causes four memory conflicts. Accessing this vector has a cost of 40 units from rule 2 above.

0	1	2	3	4
a_{11}	a_{12}	a_{13}	a_{14}	a_{15}
a_{21}	a_{22}	a_{23}	a_{24}	a_{25}
a_{31}	a_{32}	a_{33}	a_{34}	a_{35}
a_{41}	a_{42}	a_{43}	a_{44}	a_{45}
a_{51}	a_{52}	a_{53}	a_{54}	a_{55}

Figure 4-3 Conflict free access of M vector

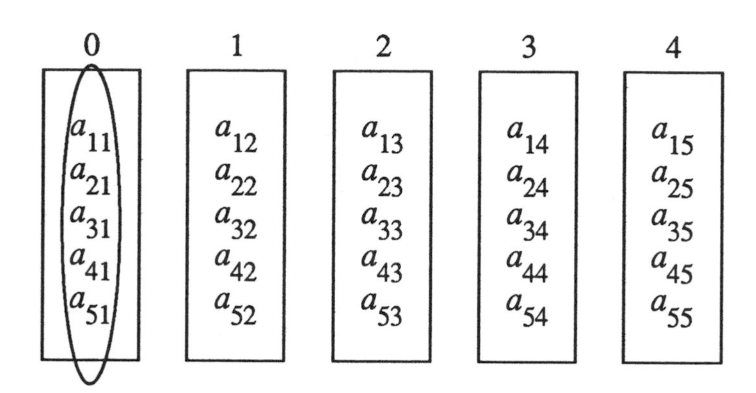

Figure 4-4 Accessing M vector with memory conflicts

Cost functions for this program segment

Figures 4-5 and 4-6 show the cost functions for accessing operands A and B just as input operations. We will consider the read operation as free except in the case where two copies of an operand are created, each in a different shape. Since operand B is only used by one operation we will only include the cost of one copy of it. However, operand A is shared, or used by two operations. The discussion in section 1-4 gave several options for finding the shape or shapes for a shared node which gives the lowest overall cost for both operations. The cost table in Figure 4-6 allows for all possibilities. The first row gives the cost for one copy of A, in row order. The second row of the table gives the cost for two copies of A. A.1, the input to the + operation, is in row order. A.2, the input to the × operation, is in column order. The cost of a transpose, or producing an operand in two shapes, is 10 units, from rule 3 above.

$S_{A.1}$	$T_B(S_B)$
row	0
col	0

Figure 4-5 Cost table for reading operand B into memory

$S_{A.1}$	$S_{A.2}$	$T_A(S_{A.1}, S_{A.2})$
row	row	0
row	col	10
col	row	10
col	col	0

Figure 4-6 Cost table for reading operand A into memory

Figure 4-7 shows the cost table for the × or inner product operation. We will assume that an inner product of A and B involves accessing the rows of A and the columns of B. In the following paragraphs we shall explain the memory organizations of operands A and B and how they affect the cost table. The number of memory conflicts that occur as a result of accessing M vectors necessary to do inner product is explained. From this we can show how each line of the cost table is derived. The shape assigned to the result, matrix C, does not influence the cost in this example. The lines of the cost table where S_c = row are identical to the lines where S_c = column.

S_C	$S_{A.2}$	S_B	$T_C(S_C, S_{A.2}, S_B)$
row	row	row	250
row	row	col	100
row	col	row	400
row	col	col	250
col	row	row	250
col	row	col	100
col	col	row	400
col	col	col	250

Figure 4-7 Cost table for inner product operation

Figure 4-9 shows the location of the elements of A and B in memory if row order was chosen for both matrices. This corresponds to the first line in the cost table in figure 4-8. Accessing the 5 rows of A corresponds to 5 conflict free M vector accesses. Each conflict free access of an M vector incurs a cost of 10, from rule 1 above. So accessing the 5 rows of A incurs a cost of 50 units. Accessing each column of B when B is stored in row order causes 4 memory conflicts (see discussion above), which gives a cost of 40 units. So accessing 5 columns of B incurs a total cost of 200 units. The 50 units for accessing A added to the 200 units for accessing B give the cost on line 1 (and line 5) of 250 units.

S_C	$S_{A.2}$	S_B	$T_C(S_C, S_{A.2}, S_B)$
	row	row	250
	row	col	100
	col	row	400
	col	col	250

Figure 4-8 First line of cost table: A and B in row order

0		1		2		3		4	
a_{11}	b_{11}	a_{12}	b_{12}	a_{13}	b_{13}	a_{14}	b_{14}	a_{15}	b_{15}
a_{21}	b_{21}	a_{22}	b_{22}	a_{23}	b_{23}	a_{24}	b_{24}	a_{25}	b_{25}
a_{31}	b_{31}	a_{32}	b_{32}	a_{33}	b_{33}	a_{34}	b_{34}	a_{35}	b_{35}
a_{41}	b_{41}	a_{42}	b_{42}	a_{43}	b_{43}	a_{44}	b_{44}	a_{45}	b_{45}
a_{51}	b_{51}	a_{52}	b_{52}	a_{53}	b_{53}	a_{54}	b_{54}	a_{55}	b_{55}

Figure 4-9 Organization of A and B in memory corresponding to Figu

Figure 4-11 shows the organization of the elements of A and B corresponding to the second line of the cost table. Each column of B is distributed over the M memory modules. Each column of B can be accessed as efficiently as each row of A, resulting in the lowest cost for the inner product operation.

S_C	$S_{A.2}$	S_B	$T_C(S_C, S_{A.2}, S_B)$
	row	row	250
	row	col	100
	col	row	400
	col	col	250

Figure 4-10 Second line of cost table with A in row and B in column

0		1		2		3		4	
a_{11}	b_{11}	a_{12}	b_{21}	a_{13}	b_{31}	a_{14}	b_{41}	a_{15}	b_{51}
a_{21}	b_{12}	a_{22}	b_{22}	a_{23}	b_{32}	a_{24}	b_{42}	a_{25}	b_{52}
a_{31}	b_{13}	a_{32}	b_{22}	a_{33}	b_{33}	a_{34}	b_{43}	a_{35}	b_{53}
a_{41}	b_{14}	a_{42}	b_{24}	a_{43}	b_{34}	a_{44}	b_{44}	a_{45}	b_{54}
a_{51}	b_{15}	a_{52}	b_{25}	a_{53}	b_{35}	a_{54}	b_{45}	a_{55}	b_{55}

Figure 4-11 Organization of A and B in memory corresponding to Figure 3-10

Figure 4-12 shows the cost table for the + operator. The reader can see that the lowest cost rows in the cost table are the rows where the shapes of the two input operands match. The shape of the result is not considered in this cost function.

$S_{A.1}$	S_C	$T_D(S_{A.1}, S_C)$
row	row	100
row	col	250
col	row	250
col	col	100

Figure 4-12 Cost table for + operation

Step one of solution (collapsible graph algorithm)

In step one of the solution, transformation A is applied to the two nodes circled in figure 4-13. Transformation A is chosen for this reduction because its requirements match the two nodes circled in figure 4-13. If the inner product node is i and the input node is j the reader can see that: outdegree(j) = 0 and node i is the only parent of node j, although node i may have other children. (see Transformation A in chapter 3) Arc B is eliminated as a result of this transformation. A shape S_B is chosen for B such that the sum of cost functions $T_C(S_C, S_{A.2}, S_B)$ and $T_B(S_B)$ is minimized, relative to S_C and $S_{A.2}$.

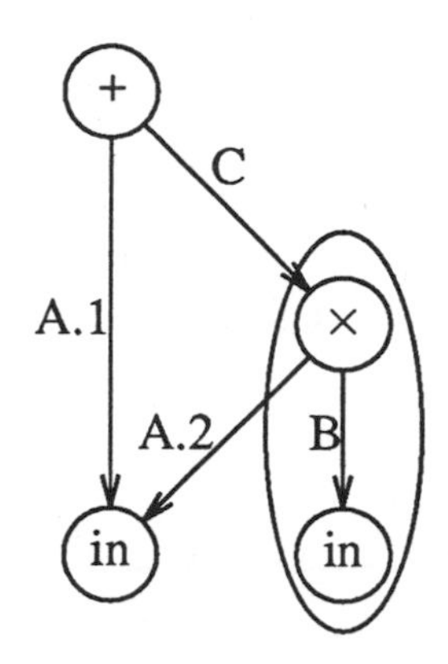

Figure 4-13 In step 1 transformation A is applied to the circled nodes

The two terms on the right hand side of the minimization equation for step one represent the cost functions of the two nodes circled in Figure 4-13. The cost function for the input node is called T_B since its result is operand B. T_B only depends on the shape S_B assigned to the arc labeled B in Figure 4-13. The cost function for the inner product node is called T_C since its result is operand C. It depends on three arguments, S_B, the shape assigned to the arc labeled B, $S_{A.2}$, the shape assigned to the arc labeled A.2, and S_C, the shape assigned to the arc labeled C. The term on the left hand side of the equation is the cost function of the new node, BC which is the result of applying transformation A. This cost function (see figure 4-14) depends on two arguments, the shape assigned to arc S_C and the shape assigned to arc $S_{A.2}$. For each combination of S_C and $S_{A.2}$ a shape is chosen for S_B which gives the minimum cost.

$$T_{BC}[S_C, S_{A.2}] = \min \{(T_C[S_C, S_{A.2}, row] + T_B[row]), \\ (T_C[S_C, S_{A.2}, col] + T_B[col])\}$$

Figure 4-14 shows the new cost table and the new graph. For each line in the new cost table, a shape for B has been chosen that minimizes the sum subject to S_C and $S_{A.2}$. This shape is shown to the left of each line in the table.

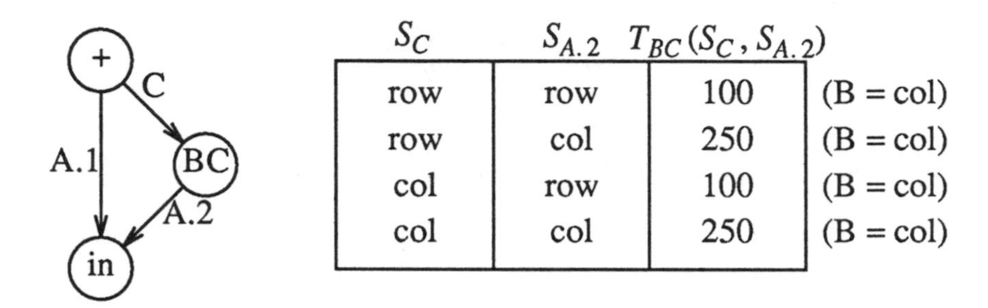

S_C	$S_{A.2}$	$T_{BC}(S_C, S_{A.2})$	
row	row	100	(B = col)
row	col	250	(B = col)
col	row	100	(B = col)
col	col	250	(B = col)

Figure 4-14 Graph and cost table after step 1

Step two of solution

In step two of the solution transformation B is applied to the two nodes circled in figure 4-15. Transformation B is chosen because the two nodes circled in figure 4-15 match the structure required for transformation B. Suppose the + node is i and the BC node is j in Case 3 of Transformation B. Indegree(i) $\leq$ 1, outdegree(i) = 1, and i is j's only parent.

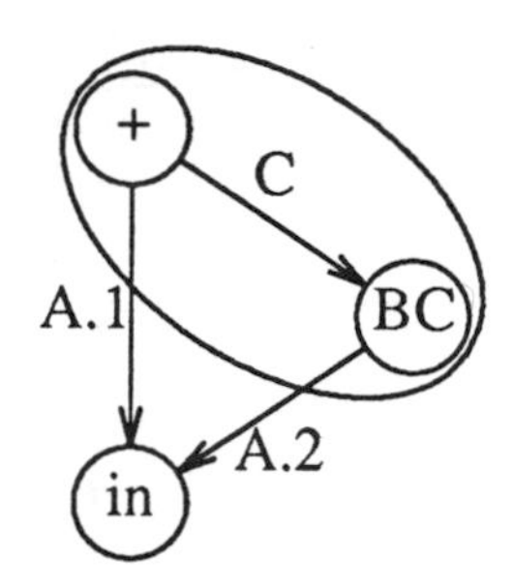

Figure 4-15 In step 2 transformation B is applied to the circled nodes

The two terms of the right hand side of the minimization equation for step 2 represent the cost functions of the nodes circled in Figure 4-15. The cost function for the + node is labeled T_D since the result of this node is operand D. This cost function depends on S_C, the shape assigned to operand C, and $S_{A.1}$, the shape assigned to the arc labeled A.1. The cost function for the BC node is labeled T_C since the result of this node is operand C. The cost function depends on S_C and $S_{A.2}$. The cost function for the new, combined node depends on $S_{A.1}$ and $S_{A.2}$.

$$T_{CD}[S_{A.1}, S_{A.2}] = \min \{(T_D[S_{A.1}, row] + T_C[row, S_{A.2}]), \\ (T_D[S_{A.1}, col] + T_C[col, S_{A.2}])\}$$

Arc C is eliminated. A shape S_C is chosen to minimize the sum of $T_C(S_C, S_{A.2})$ and $T_D(S_D, S_{A.2}, S_C)$. Figure 4-16 shows the cost tables for $T_{BC}(S_C, S_{A.2})$ (left) and $T_+(S_{A.2}, S_C)$ (right). The rows of the table that gave the minimum cost sum are circled.

S_C	$S_{A.2}$	$T_{BC}(S_C, S_{A.2})$
row	row	100
row	col	250
col	row	100
col	col	250

$S_{A.1}$	S_C	$T_+(S_{A.1}, S_C)$
row	row	100
row	col	250
col	row	250
col	col	100

Figure 4-16 Step 2 - the sum of the circled rows gives the minimum cost

Finally, Figure 4-17 shows the new cost table and the new graph. The shape for C which gives the minimum total cost, where $S_{A.1}$ and $S_{A.2}$ are the shapes shown in this table, is displayed to the right of each line in the table.

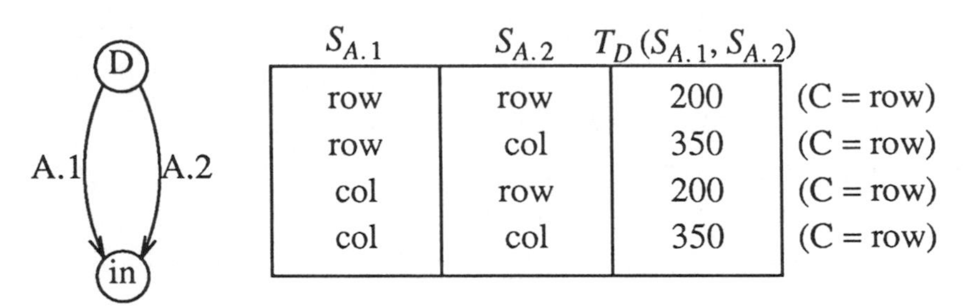

$S_{A.1}$	$S_{A.2}$	$T_D(S_{A.1}, S_{A.2})$	
row	row	200	(C = row)
row	col	350	(C = row)
col	row	200	(C = row)
col	col	350	(C = row)

Figure 4-17 Graph and cost table after step 2

Step 3

In step 3 of the solution, transformation A is applied to the only two nodes remaining in the graph. (see Figure 4-18)

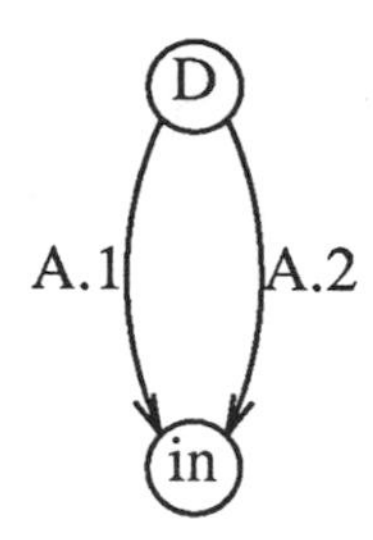

Figure 4-18 In step 3 transformation A is applied to the two remaining nodes

The minimization function follows the same pattern as the two previous ones.

$$\begin{aligned} T_{DA}[S_D] = \min\{ & (T_D[S_{A.1}^{row}, S_{A.2}^{row}] + T_A[S_{A.1}^{row}, S_{A.2}^{row}]), \\ & (T_D[S_{A.1}^{col}, S_{A.2}^{row}] + T_A[S_{A.1}^{col}, S_{A.2}^{row}]), \\ & (T_D[S_{A.1}^{row}, S_{A.2}^{col}] + T_A[S_{A.1}^{row}, S_{A.2}^{col}]), \\ & (T_D[S_{A.1}^{col}, S_{A.2}^{col}] + T_A[S_{A.1}^{col}, S_{A.2}^{col}])\} \end{aligned}$$

$S_{A.1}$	$S_{A.2}$	$T_D(S_{A.1}, S_{A.2})$
row	row	200
row	col	350
col	row	200
col	col	350

$S_{A.1}$	$S_{A.2}$	$T_A(S_{A.1}, S_{A.2})$
row	row	0
row	col	10
col	row	10
col	col	0

Figure 4-19 Step 3 - the circled rows give the minimum cost

Arcs A.1 and A.2 are eliminated. Shapes $S_{A.1}$ and $S_{A.2}$ are chosen to minimize the sum of $T_D(S_D, S_{A.1}, S_{A.2})$ and $T_A(S_{A.1}, S_{A.2})$. The shapes for A.1 and A.2 that minimize this function are both row order.

Final result

The final minimum cost for this graph is 200 units. The assignment of shapes to arcs that gives this cost is:

S_B = column

S_C = row

$S_{A.1}$ = row

$S_{A.2}$ = row

Chapter 5

SHAPES PROBLEM COMPLEXITY ISSUES

In this chapter we give evidence for the NP-completeness of the shapes problem. The chapter contains three separate proofs, each of which is a reduction from a known NP-complete problem. Two of these proofs are reductions from set cover, and will be known as set cover I, and set cover II. The third proof is a reduction from the 3-colorability problem.[4, 10, 11, 13, 14]

The three proofs apply to versions of the shapes problem in which different parameters of the problem are restricted. In other words, in each example, some variables of the problem are constrained, while others are relaxed. In a graph, a shared node is a node with more than one parent. The number of shared nodes in the graph is an important characteristic of the shapes problem. For example, in set cover I the number of shared nodes is restricted to the constant 1, but the number of shapes is a variable as is the number of node "types" (operator symbols). In set cover II the number of shared nodes is restricted to 1 and the number of table types is restricted to the constant 4 but the number of shapes is still a variable. In 3-colorability, the number of shared nodes is allowed to be a variable but the number of table types and the number of shapes are restricted to constants. The intent of these reductions is to show that the unrestricted shapes problem is NP-complete by showing that many restricted classes of the shapes problem which are apparently simpler than the original problem are still NP-complete.

Definition of the shapes problem

In the preceding chapters we gave a procedure for solving the shapes problem, and demonstrated how these procedures could be applied. In this section we bring the elements of the problem together into a precise definition.

Definition: Let a graph $G = (n, a)$ where n is a set of nodes and $a \subset n \times n$ is a set of directed arcs. Each arc $\alpha \subset a$ can take any shape in the set of shapes S_{α}. Each node has an associated cost function. This cost function depends on the node's *operator label* e.g. +, ×, transpose, etc, and on the shapes assigned to all arcs which touch that node. A *configuration* is an assignment of shapes to arcs for an entire graph. The cost of a given configuration is the sum of the cost functions of all the nodes in the graph. The *shapes problem* is to find a configuration of minimum cost.

The reader should note that three variables in the shapes problem definition play a major role in determining how difficult the problem is. These variables are: the number of shapes, the number of different types of cost functions (operator types), and the number of shared nodes.

The job of transforming another problem into an equivalent shapes problem can be made easier by simplifying the shapes problem definition. We will change the problem definition from a minimization problem to a yes/no problem. We can do this by changing the last line of the definition above to "Is there a configuration of cost ≤ k, for some given number k." This does not change the overall complexity of the shapes problem for the following reason. If the answer to this yes/no formulation can be found in polynomial time for every possible k, the minimum value for k can be easily found in polynomial time by doing a binary search over all possible values of k.[4]

The shapes problem is in NP

A nondeterministic Turing machine can first guess a configuration of shapes to assign to all arcs. The Turing machine can then *in parallel* test all guesses to see that the sum of all node cost function values is equal to k for at least one of the guessed configurations. As far as individual cost functions are concerned, if we assume that each operation accepts only two operands and produces one result, and the number of possible shapes is s, a table of size s^3 suffices to store the costs, allowing a nondeterministic Turing machine to compute the node's cost function in constant time.

Definition of set cover:

The first two reductions in this chapter use the set cover problem. This problem is known to be NP-complete.[4] The set cover problem can be stated as:

Given a family of sets $(A_1 \cdots A_n)$ and the integer k, does there exist a subfamily of k sets $(A_{i_1} ... A_{i_k})$ such that

$$\bigcup (A_{i_1} \cdots A_{i_k}) = \bigcup (A_1 ... A_k).$$

Cost functions for shared nodes in the set cover proofs

The cost of each shared node will be taken to be the number of distinct shapes assigned to arcs which touch the node, minus 1. This function was chosen to be easily computed from a configuration, and to be specifiable without requiring a table whose size grows exponentially with the number of arcs touching a node. An example of a subroutine for computing such a function can be found in Appendix F. This cost function also has meaning for the matrix application. Suppose a matrix was shared by several operations, and one alternative course of action was to make different copies of the matrix in different shapes. One of these copies would be present as an input. The others would be created by transpositions. So to have a matrix in m shapes would require m - 1 transpositions, or the number of different shapes - 1.

Set cover I

Theorem 4.1: The set cover problem is polynomially transformable to the shapes problem with the number of shared nodes restricted to one. Therefore, this restricted version of the shapes problem is NP-complete.

Proof: Let S be a set cover instance with a family of n sets $(A_1 \cdots A_n)$ made up of u elements $(e_1 \cdots e_u)$. Let k be an integer less than n. We shall construct a shapes problem P consisting of a directed acyclic graph D with a cost function for each node.

Graph structure

D consists of a chain of nodes and a single shared node. For each element e_i in the set cover instance there is a node, also called e_i, in the chain. Each of the u nodes in the chain has the shared node as one child and the next node down in the chain as the other child.

Each of the u nodes in the chain is labeled with one of the u elements in the set cover instance. The shared node is labeled *node 1*. Each arc in the graph is labeled with the name of the node immediately below it. For example the arc which is the parent of a node labeled e_i is called arc e_i. The shared node has u parent arcs, one for each node in the chain. For each node e_i in the chain the arc from it to the shared node is labeled $1.e_i$.

Each set $(A_1 \cdots A_n)$ in the set cover instance corresponds to a shape which can be assigned to the shared node arcs. These shapes will be called $(A_1 \cdots A_n)$ also.

Cost functions

Each of the u nodes in the chain has a different cost table associated with it. The cost for a node labeled e_i depends only on the shape assigned to the shared node arc labeled $1.e_i$. In other words, if some shape is assigned to arc $1.e_i$ the cost is the same whatever the shapes assigned to other arcs touching node e_i. The

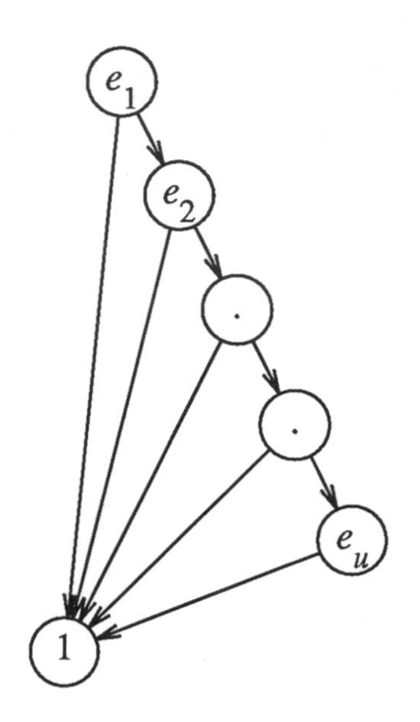

Figure 5-1 Set cover graph

cost table for each node e_i contains n entries, one for each shape $(A_1 \cdots A_n)$ which could be assigned to arc $1.e_i$. The cost of node e_i is 0 if the shape A_j assigned to arc $1.e_i$ corresponds to a set A_j in the set cover instance such that set A_j contains element e_i. The cost for node e_i is n otherwise. Call the cost table for element e_i table T_{e_i}. The term for the shape on arc $1.e_i$ is $S_{1.e_i}$. The shapes which are assigned to the shared node arcs are the variables of the problem. Assume shape A_j is assigned to arc $1.e_i$.

$$T_{e_i}[S_{1.e_i}] = 0 \text{ if } \textit{element } e_i \in \textit{set } A_j$$

$$= n \textit{ otherwise}$$

The cost of the shared node is the number of distinct shapes assigned to shared node arcs - 1. The cost for the entire graph is,

$$\sum_{i=1}^{u} T_{e_i}[S_{1.e_i}] + number\ of\ shapes\ assigned\ shared\ node\ arcs - 1$$

Time required for reduction

The above transformation from a set cover instance to a shapes instance takes time which is a polynomial function of the size of the set cover problem. The set cover problem has n sets made up of u elements. The graph D has u chain nodes and one shared node. Each chain node's cost function is described as a table with exactly n entries. The shared node's cost is computed by counting the number of shapes assigned to its nodes. This number can be no greater than n - 1.

We shall show that for every $k < n$, P has a configuration of cost k - 1 if and only if S has a subfamily of size k which covers. Note that P always has a configuration of cost n - 1 and S always has a subfamily of size n, trivially.

If

Assume that S has a subfamily of size k which covers.

Let $(A_{j_1} \cdots A_{j_k})$ be a subfamily of sets which covers, that is every element $(e_1 \cdots e_u)$ is a member of at least one of $(A_{j_1} \cdots A_{j_k})$. Assign the shapes which correspond to sets $(A_{j_1} \cdots A_{j_k})$ to the arcs touching the shared node. This assignment is made such that for each arc $1.e_i$, a shape A_{e_i} is chosen from $(A_{j_1} \cdots A_{j_k})$ such that element $e_i \in$ set A_{e_i}. Since the subfamily is a cover set, for each e_i, an A_{e_i} can always be found. The cost of each chain node e_i is 0, because for each node e_i and each shape

A_{e_i} chosen, element $e_i \in A_{e_i}$. The cost of the shared node is the number of distinct shapes assigned to its arcs minus one, or k-1. Therefore, P has a configuration of cost k-1.

Only if

Assume that P has a configuration of cost k-1 < n. If k-1 < n, all the chain nodes must have cost = 0, since the smallest non-zero cost a chain node can have is n. Since all the chain nodes have cost = 0, the subfamily of sets corresponding to the shapes assigned to the shared node arcs must be a cover set. Also, since all the chain nodes have cost = 0, the cost for the rest of the configuration equals the cost of the shared node alone. If the cost of the shared node is k-1 there must be k distinct shapes assigned to the shared node arcs. Therefore, there are k sets in the subfamily which corresponds to the set of shapes assigned to shared node arcs. Therefore, S has a subfamily of k sets which covers.

Example of set cover I.

The elements are {a,b,c,d}, the sets are {(a,b), (a,c), (b,c), (c,d)}, and the integer k is 2 in the set cover instance S.

The graph D, illustrated in Figure 5-2, is created.

Each node in the chain is labeled with one of the elements. Each of these has a different cost table. The cost tables for each node are shown in Table 5-1.

The number of distinct shapes - 1 is the shared node's cost function.

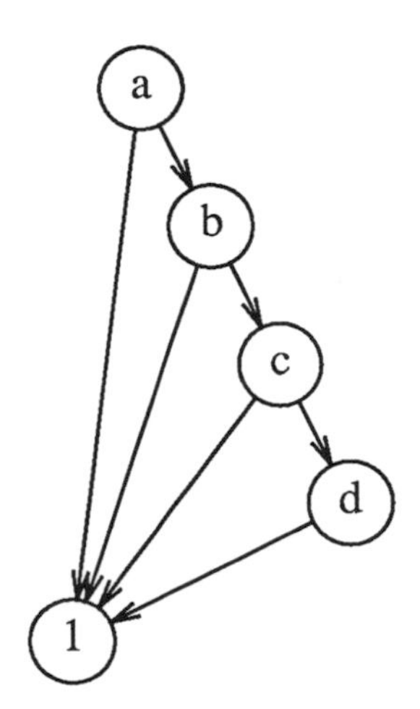

Figure 5-2 Shapes problem constructed from set cover instance

$S_{1.a}$	$T_a[S_{1.a}]$
(a,b)	0
(a,c)	0
(b,c)	n
(c,d)	n

$S_{1.c}$	$T_c[S_{1.c}]$
(a,b)	0
(a,c)	n
(b,c)	0
(c,d)	n

$S_{1.b}$	$T_b[S_{1.b}]$
(a,b)	n
(a,c)	0
(b,c)	0
(c,d)	0

$S_{1.d}$	$T_d[S_{1.d}]$
(a,b)	n
(a,c)	n
(b,c)	n
(c,d)	0

Table 5-1 Cost tables for set cover I

The cost for the configuration is

$$\sum_{i=1}^{i=u} T_{e_i}[S_{1.e_i}] + number\ \ of\ \ distinct\ \ shapes - 1$$

shape subset	sum of T's	#shapes - 1	total cost
(a,b)	2n = 8	0	8
(a,c)	2n = 8	0	8
(b,c)	2n = 8	0	8
(c,d)	2n = 8	0	8
(a,b)(a,c)	n = 4	1	5
(a,b)(b,c)	n = 4	1	5
(a,b)(c,d)	0	1	1
(a,c)(b,c)	n = 4	1	5
(a,c)(c,d)	n = 4	1	5
(b,c)(c,d)	n = 4	1	5
(a,b)(a,c)(b,c)	n = 4	2	6
(a,b)(a,c)(c,d)	0	2	2
(a,b)(b,c)(c,d)	0	2	2
(a,c)(b,c)(c,d)	0	2	2

Table 5-2 Cost table for set cover I

The shapes corresponding to sets (a,b) and (c,d) are the ones which give the cost of 1 for the configuration. It is easily seen that (a,b) and (c,d) are a cover set of size 2 for this instance of set cover.

Set cover II

In set cover I the number of types of operator nodes, or types of cost tables in the shapes instance is a variable in the set cover instance. In set cover I it is u, the number of elements. It would be possible, as in the example of set cover I for each operator node in D to be a different type. One could argue that if D were a real program graph, that the number of types of operator nodes would probably be a constant. Forcing the number of types of operator nodes to be a constant is also a simplification of the problem. Exploring whether a simplification of an NP-complete problem has a polynomial solution or is also NP-complete is an interesting issue.[10]

Set cover II imposes the following restriction on D. The number of types of operator nodes or types of cost tables is restricted to the constant 4. The shapes problem remains NP-complete despite this simplification. As in set cover I, D is restricted to a single shared node. As far as other parameters of the problem are concerned, the number of shapes is a variable in the problem, as is the number of nodes in D.

Theorem 4.2: The set cover problem is polynomially transformable to the shapes problem with the number of types of operator nodes restricted to a constant, and the number of shared nodes restricted to one.

Proof: Let S be a set cover instance with a family of n sets $(A_1 \cdots A_n)$ made up of u elements $(e_1 \cdots e_u)$. Let k be an integer less than n. We shall construct a shapes problem P consisting of a directed acyclic graph D with a cost function for each node.

Graph structure and cost functions

D is a directed acyclic graph with a single shared node. For each element e_i in the set cover instance there is a subgraph, also called e_i, in D. The subgraph consists of a chain of nodes. The bottom node in each subgraph is a direct parent of the shared node (see Figure 5-6).

The sets $(A_1 \cdots A_n)$ in the set cover instance correspond to the shapes, which will also be called $(A_1 \cdots A_n)$. There is also a *yes* shape which all arcs except shared node arcs can take. The shapes $(A_1 \cdots A_n)$ are given a linear ordering such that $A_{j-1} < A_j$ in the ordering. Assume that i is a node, T_i is its cost table, S_i is the shape assigned to the arc above i and S_{i-1} is the shape assigned to the arc below i. The nodes in the chain are numbered from the bottom up. The subscript j of shape A_j indicates its position in the ordering: shape A_{j-1} precedes shape A_j. There are four types of nodes, or cost tables. In each case we will illustrate the cost table with a figure.

The cycle node has the following table:

$$
\begin{aligned}
T_i[S_i, S_{i-1}] &= 0 \text{ if } S_{i-1} \textit{ is shape } A_{j-1}, S_i \textit{ is shape } A_j \\
&= 0 \text{ if } S_{i-1} \textit{ is shape } A_n, S_i \textit{ is shape } A_1 \\
&= 0 \text{ if } S_{i-1} \textit{ is shape yes}, S_i \textit{ is shape yes} \\
&= n \textit{ otherwise}
\end{aligned}
$$

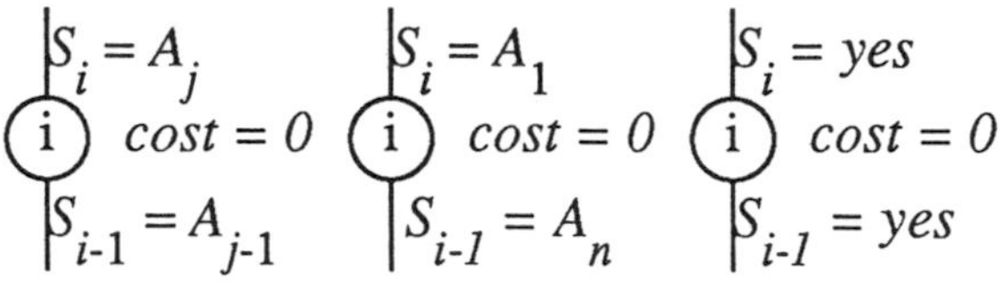

$S_i = A_j$ (i) *cost = 0* $S_{i-1} = A_{j-1}$

$S_i = A_1$ (i) *cost = 0* $S_{i-1} = A_n$

$S_i = yes$ (i) *cost = 0* $S_{i-1} = yes$

Figure 5-3 Shape assignments that give a cycle node cost = 0

The cost table for the recognize node is the same as the cycle node except for one difference. The cycle node has cost = 0 if shape A_n is converted to shape A_1, a cyclic pattern, whereas the recognize node has cost = 0 if shape A_n is converted to shape *yes*. The recognize node has the following table:

$$T_i[S_i, S_{i-1}] = 0 \text{ if } S_{i-1} \textit{ is shape } A_{j-1}, S_i \textit{ is shape } A_j$$

$$= 0 \text{ if } S_{i-1} \textit{ is shape } A_n, S_i \textit{ is shape yes}$$

$$= 0 \text{ if } S_{i-1} \textit{ is shape yes}, S_i \textit{ is shape yes}$$

$$= n \textit{ otherwise}$$

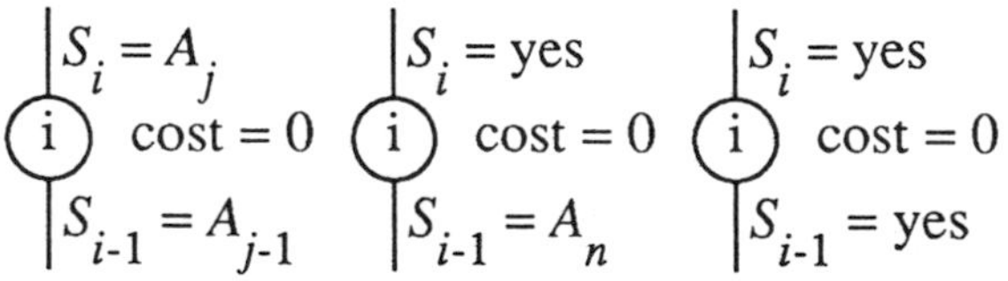

$S_i = A_j$ (i) cost = 0 $S_{i-1} = A_{j-1}$

S_i = yes (i) cost = 0 $S_{i-1} = A_n$

S_i = yes (i) cost = 0 S_{i-1} = yes

Figure 5-4 Shape assignments that give a recognize node cost = 0

The yes/no node only has an input shape. The yes/no node

has the following table:

$$T_i[S_{i-1}] = 0 \textit{ if } S_{i-1} \textit{ is shape yes}$$
$$= n \textit{ otherwise}$$

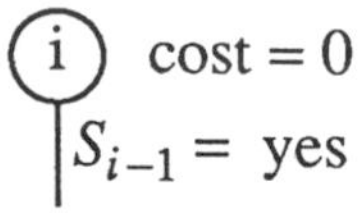

(i) cost = 0

S_{i-1} = yes

Figure 5-5 Shape assignment that gives a yes/no node cost = 0

The shared node has the same cost function as in set cover I. This cost is the number of distinct shapes assigned to arcs which touch the shared node - 1.

Suppose out of n sets $(A_1 \cdots A_n)$ element e_i is contained in m sets $(A_{j_1} \cdots A_{j_m})$. We will construct a subgraph e_i to correspond to element e_i. Each subgraph contains one shared node arc. The shared node is the bottom node in Figure 5-6. The subgraph corresponding to element e_i is constructed from cycle and recognize nodes in such a way that it will have total cost = 0 if and only if the shape assigned to its shared node arc is one of $A_{j1} \cdots A_{jm}$.

Assume $j_1 < j_2 < \cdots j_m$ in the linear ordering. The total number of nodes in the subgraph, not including the shared node, is n. Each shape A_{j_1} in $(A_{j_1} \cdots A_{j_m})$ has one recognize node at position $n - j_i + 1$ in the subgraph, so there are exactly m recognize nodes in the subgraph. All other nodes in the subgraph, except for the root node, are cycle nodes. The top recognize node is at position $n - j_1 + 1$ in the subgraph. Above this node is a single

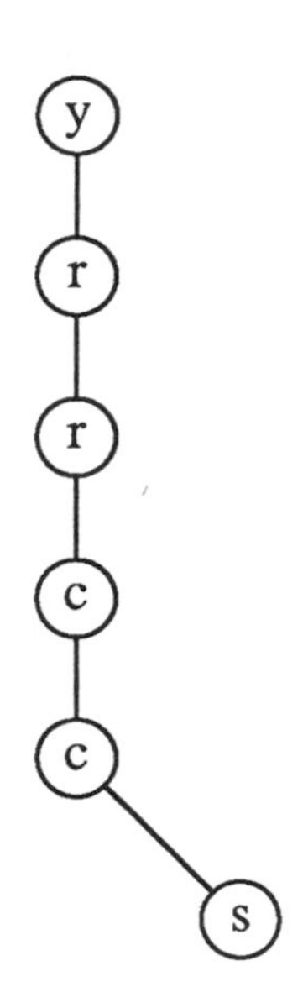

Figure 5-6 Subgraph corresponding to element a

yes/no node which is at the root of the subgraph. This subgraph e_i is constructed so that each of the nodes in the subgraph has cost = 0 when one of the shapes corresponding to sets that contain e_i is assigned to the shared node arc.

Suppose that shape A_p is assigned to the shared node arc of subgraph e_i such that element e_i is a member of set A_p. The bottom section of subgraph e_i consists of a chain of $n-p$ cycle and recognize nodes. There is one assignment of shapes to the arcs of subgraph e_i which will make every node of the subgraph have cost = 0. Call the bottom node in subgraph e_i node 1, and assume that shape A_p is assigned to e_i's shared node arc. Node 1 has cost = 0 if shape A_p is assigned to the shared node arc and shape $A_p + 1$ is assigned to arc 1. The second node up the chain will be called node 2. If shape $A_p + 1$ is assigned to arc 1, then node 2 has cost = 0 if shape $A_p + 2$ is assigned to arc 2. The n-p'th node up the chain will be called node n-p. If shape $A_{p+n-p-1}$, or A_{n-1} is assigned to arc n-p-1, then node n-p has cost = 0 if shape A_{p+n-p} or A_n is assigned to arc n-p. Subgraph e_i was constructed such that if element ei is a member of set A_p, then the node at position n-p+1 is assigned the *yes* shape. If all arcs in e_i above node n-p+1 are assigned the *yes* shape all cycle and recognize nodes above n-p+1 will have cost = 0, and and the yes/no node at the top of the subgraph will have cost = 0.

Time requirements for the reduction

The above transformation from a set cover instance S to a shapes instance P takes time which is a polynomial function of the size of the set cover instance. The number of cycle and recognize nodes in each subgraph is less than or equal to the number of sets n, plus 1. The number of subgraphs is equal to u, the number of elements in the union of all the sets.

The number of entries in the cycle and recognize nodes' tables is equal to the square of the number of sets $n + 1$. The number of entries in the yes/no nodes tables is equal to $n + 1$.

We shall show that for all $k < n$, P has a configuration of cost k - 1 if and only if S has a subfamily of size k that covers.

If

Assume S has a subfamily $(A_{j_1} \cdots A_{j_k})$ of size k that covers. Assign the corresponding set of shapes $(A_{j_1} \cdots A_{j_k})$ to the shared node arcs as follows. For each subgraph e_i, assign the shared node arc e_i a shape A_p chosen from $(A_{j_1} \cdots A_{j_k})$ such that element $e_i \in$ set A_p. Since $(A_{j_1} \cdots A_{j_k})$ is a cover set this can always be done. For shape A_p, p is its index in the linear ordering. Since A_p is one of the sets that contains element e_i, the subgraph e_i has been constructed such that node n-p+1 from the bottom is a recognize node. For shape A_p there are n-p shapes remaining in the sequence. For the lower n-p nodes in subgraph e_i assign shapes as follows. The shared node arc is assigned shape A_p. For each of the n-p lower nodes, if shape A_i is assigned to a node's child arc, assign shape A_{i+1} to its parent arc. Assign the parent arc of the recognize node n-p+1 shape *yes*, and assign this shape to all arcs above the recognize node in e_i but below e_i's y/n node. These assignments can always be done, and they assure that all node costs are zero, since shape A_p assigned to the bottom arc causes shape $A_{p+(n-p)}$ or A_n to be assigned to the child arc of the recognize node. Since the result of the top recognize node in the chain is assigned the yes shape, the y/n node will have cost=0. Since each subgraph e_i has cost=0, the cost for the configuration equals the cost of the shared node. This cost is the number of distinct shapes assigned to the shared node minus 1. Therefore, P has a cost of k - 1.

Only if

Assume P has a configuration of cost k - 1. Since the cost of the configuration is less than n, the cost of each subgraph e_i must be 0, since the smallest non-zero cost a subgraph e_i can have is n. Since the cost of each subgraph e_i is 0, the shape A_p assigned to each shared node arc e_i corresponds to a set A_p such that element e_i is in set A_p. If this is true for each subgraph e_i, then the subfamily of sets corresponding to the shapes assigned to the shared node arcs must cover. (see above) Since each subgraph has cost equal to 0, the cost for the configuration equals the cost of the shared node alone. If the cost of the shared node is k - 1, the number of shapes assigned to arcs of the shared node is k. So the set of k shapes assigned to shared node arcs corresponds to a subfamily of k sets. Therefore, S has a subfamily of size k that covers.

Example of set cover II

The set cover instance is the same as in set cover I. {a,b,c,d} are the elements and {(a,b), (a,c), (b,c), (c,d)} are the sets. Each set corresponds to a shape. We will call the shapes A_1, A_2, A_3, A_4 and they will correspond to the sets as follows. A_1 corresponds to (a,b), A_2 corresponds to (a,c), A_3 corresponds to (b,c), and A_4 corresponds to (c,d). As far as the linear ordering of the shapes is concerned, $\{A_1, A_2, A_3, A_4\}$ is the order.

The directed acyclic graph illustrated in Figure 5-7 is constructed. Each of the four subgraphs corresponds to one element of the set cover instance. The element corresponding to each subgraph is shown above the subgraph in the figure.

Subgraph a will have cost equal to 0 if and only if either the shape corresponding to set (a,b) or the shape corresponding to set (a,c) is assigned to the shared node arc. In Figure 5-8 we show the subtree for element a. In each of the 4 cases, the shared node arc is assigned a different shape. We will call shape (a,b) A_1, shape (a,c) A_2, shape (b,c) A_3, and shape (c,d) A_4.

As Figure 5-8 shows, the subgraph for element a can only have cost equal to 0 if either A_1, the shape corresponding to set (a,b), or A_2, the shape corresponding to set (a,c) is assigned to the

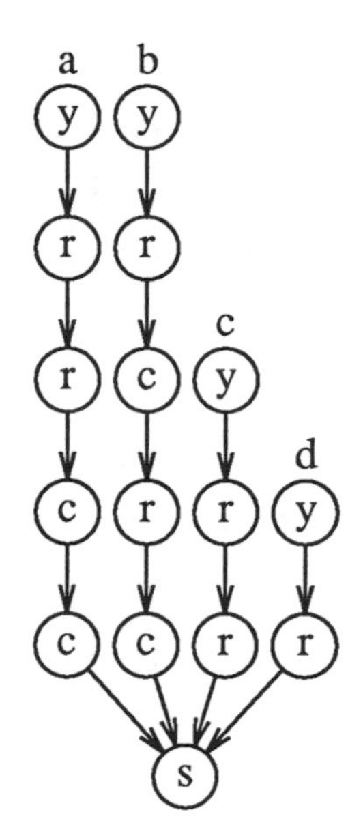

Figure 5-7 Shapes graph constructed from set cover II
shared node arc.

The following are examples of cost tables for cycle, recognize, and yes/no nodes for this instance. In each table, the input shape corresponds to the arc below the node, and the output shape corresponds to the arc above the node.

The cycle node converts each input shape to the output shape which is next in the linear ordering. If the input shape is the last one in the linear ordering, the output shape will be the first shape in the linear ordering. If the input shape is *yes*, the output shape will be *yes* also.

The recognize node, shown in Table 5-11, behaves like the cycle node in that it converts each input shape to the output shape which is next in the linear ordering. The difference is that the recognize node converts the last shape in the linear ordering to the *yes* shape. As in the case of the cycle node, if the input shape is

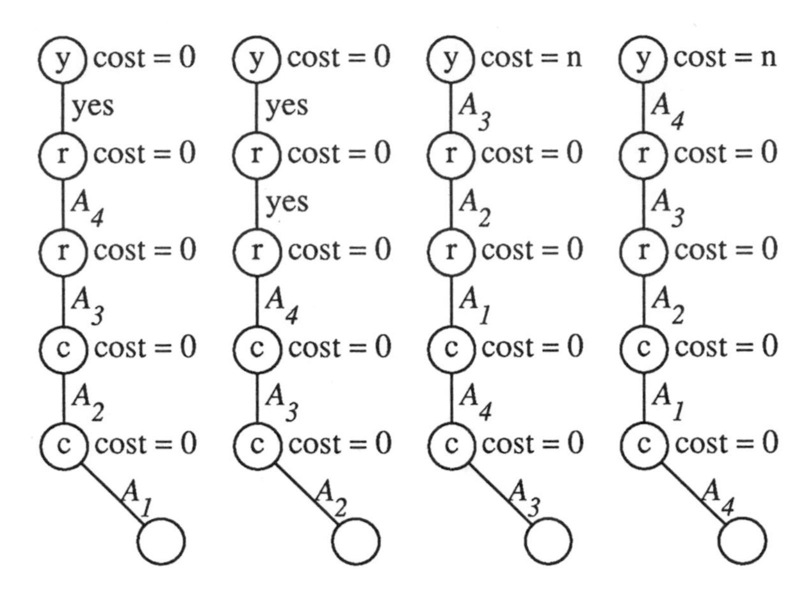

Figure 5-8 Subgraph corresponding to element a, with different shapes assigned to the shared node arc

yes the output shape will also be *yes*.

The yes/no node, illustrated in Table 5-12, only has an input shape. These nodes are at the very top of the graph. The yes/no node can only have a cost of 0 if the input shape is *yes*.

input shape	result shape	cost
(a,b)	(a,b)	n
(a,b)	(a,c)	0
(a,b)	(b,c)	n
(a,b)	(c,d)	n
(a,b)	yes	n
(a,c)	(a,b)	n
(a,c)	(a,c)	n
(a,c)	(b,c)	0
(a,c)	(c,d)	n
(a,c)	yes	n
(b,c)	(a,b)	n
(b,c)	(a,c)	n
(b,c)	(b,c)	n
(b,c)	(c,d)	0
(b,c)	yes	n
(c,d)	(a,b)	0
(c,d)	(a,c)	n
(c,d)	(b,c)	n
(c,d)	(c,d)	n
(c,d)	yes	n
yes	(a,b)	n
yes	(a,c)	n
yes	(b,c)	n
yes	(c,d)	n
yes	yes	0

Table 5-10 Cost table for cycle node

input shape	result shape	cost
(a,b)	(a,b)	n
(a,b)	(a,c)	0
(a,b)	(b,c)	n
(a,b)	(c,d)	n
(a,b)	yes	n
(a,c)	(a,b)	n
(a,c)	(a,c)	n
(a,c)	(b,c)	0
(a,c)	(c,d)	n
(a,c)	yes	n
(b,c)	(a,b)	n
(b,c)	(a,c)	n
(b,c)	(b,c)	n
(b,c)	(c,d)	0
(b,c)	yes	n
(c,d)	(a,b)	n
(c,d)	(a,c)	n
(c,d)	(b,c)	n
(c,d)	(c,d)	n
(c,d)	yes	0
yes	(a,b)	n
yes	(a,c)	n
yes	(b,c)	n
yes	(c,d)	n
yes	yes	0

Table 5-11 Cost table for recognize node

input shape	cost
(a,b)	n
(a,c)	n
(b,c)	n
(c,d)	n
yes	0

Table 5-12 Cost table for yes/no node

The costs for the entire graph are the same as the example in set cover I, so the table is not repeated here. The set of shapes which give a cost of 1 when assigned to arcs which touch the shared node are the shapes corresponding to sets (a,b) and (c,d). These sets form a cover set of size 2 for the set cover instance.

Reduction to the 3-Colorability Problem

In this section, a third restricted version of the shapes problem is shown to be NP-complete. The proof is a reduction of the 3-colorability problem to a version of the shapes problem in which the number of shared nodes is a variable, the number of different operator node types is restricted to a constant, and the number of shapes is restricted to a constant. The reduction is from 3-colorability to 3-shapes, a version of the shapes problem in which only three shapes exist.

Definition of 3-colorability

A graph G is 3-colorable if there exists an assignment of 3 colors to the nodes of G such that no two adjacent nodes are assigned the same color. The 3-colorability problem is to determine whether a given undirected graph is 3-colorable.[4]

Construction of a shapes instance from a 3-colorability instance

Theorem 4.3: The 3-colorability problem is polynomially transformable to the shapes problem with the number of shapes restricted to a constant, and the number of operator types also a constant. Therefore, this restricted version of the shapes problem is NP-complete.

Proof: Let C be an instance of the 3-colorability problem. Let G be a graph with n nodes. Each node n_i in G is adjacent to a certain number m_i of other nodes. We shall construct an instance P of the shapes problem consisting of a directed acyclic graph D (see example). The 3 colors correspond to 3 shapes which can be assigned to arcs of D. D consists of two types of nodes, or cost tables. These will be called the = node and the ≠ node.

The cost table for the = node is:

$$T_i[S_i, S_{i-1}, S_j] = 0 \text{ if } S_i = S_{i-1} = S_j$$
$$= n \textit{ otherwise}$$

The ≠ node has two parent arcs. For node i they will be called $i.1$ and $i.2$ and the shapes assigned to these arcs will be $S_{i.1}$ and $S_{i.2}$. The cost table for the ≠ node is

$$T_i[S_{i.1}, S_{i.2}] = 0 \text{ if } S_{i.1} \neq S_{i.2}$$
$$= n \text{ if } S_{i.1} = S_{i.2}$$

For each arc (i, j) in G, there is a ≠ node, also called N_{ij}, in D. For each node n_i in G there is a chain of m_{i-1} = nodes in D.

The chain of = nodes corresponding to node n_i, and all arcs which touch those nodes are called subgraph n_i

Each ≠ node is shared by two parents. Suppose arc (i, j) in G has nodes n_i and n_j. at either end. One parent of the corresponding ≠ node N_{ij} in D is an = node in subgraph n_i. The other parent of ≠ node N_{ij} is an = node in subgraph n_j.

Each = node in subgraph n_i in D has two children. One child is a ≠ node N_{ij} corresponding to an arc in G which touches node n_i. The other child is the next = node down in the chain, if there is one. If not, the other child is another ≠ node N_{ik}, corresponding to another arc in G which touches n_i.

Time requirements for reduction

The above transformation from a colorability instance C to a shapes instance P takes time which is a polynomial function of the size of the colorability instance. The number of ≠ nodes in D is equal to the number of arcs in G (less than the number of nodes in G squared). The number of = nodes in D is less than or equal to the number of nodes in G squared.

The number of entries in an = node is equal to the number of colors: 3^3 or 27. The number of entries in the = node's table is less than or equal to the number of colors: 3^2 or 9.

We shall show that there is a configuration of cost 0 for P if and only if the graph G is 3-colorable.

If

Assume that G is 3-colorable.

For each node n_i in G, assign all the arcs in the corresponding subgraph n_i in D the shape which corresponds to the color assigned to node n_i. If every arc in subgraph n_i has the same shape, every = node in subgraph n_i will have cost = 0.

Each ≠ node N_{ij} represents the adjacency of nodes n_i and n_j in G by having one parent in subgraph n_i and one parent in subgraph n_j in D.

All arcs in subgraph n_i are the shape which corresponds to the color assigned to node n_i. All the arcs in subgraph n_j are the shape which corresponds to the color assigned to node n_j. Since n_i and n_j must have different colors, the arcs of subgraphs n_i and n_j have different shapes.

Since its two parent arcs have different shapes, node N_{ij} has cost=0. Therefore, P has a configuration of cost = 0.

Only if

Assume that D has a configuration of cost 0

Every = node in D has cost=0, so for each subgraph n_i every arc in the subgraph must be assigned the same shape. Every ≠ node in D has cost=0, so for each ≠ node N_{ij} the parent arc from subgraph n_i must be assigned a different shape than the parent arc from subgraph n_j.

Since ≠ node N_{ij} corresponds to an arc between nodes n_i and n_j, if subgraph n_i and n_j's arcs are different shapes, then nodes n_i and n_j can be different colors.

If, for each pair of nodes n_i and n_j that are connected by arc (i, j), these nodes are different colors, then G is 3-colorable.

Example of colorability reduction.

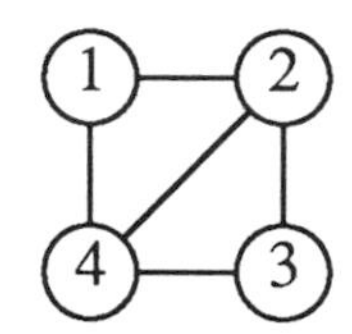

Figure 5-9 Colorability graph

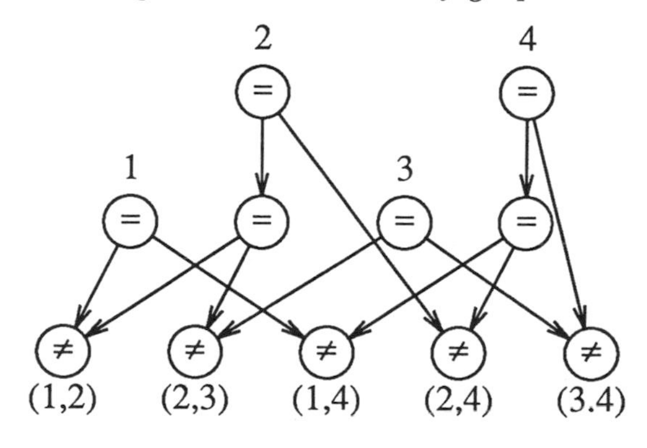

Figure 5-10 Shapes Graph constructed from Colorability Graph

Chapter 6

SHAPES SOLUTION FOR JACOBI ITERATION

In this chapter we will attempt to apply the techniques demonstrated in the earlier chapters to a real problem using a real machine. The example problem we have chosen is the well known mathematical technique of Jacobi Iteration. The example machine architecture we have chosen is the CDC Cyber 205.

Jacobi iteration was chosen as an example for three reasons. First, we wanted to choose a problem whose basic operands are matrices. Jacobi iteration primarily consists of matrix vector multiplication. Second, the effects of reshaping the matrix in matrix vector multiplication in order to make use of vector operations has been studied.[19] Investigating Jacobi iteration allows us to take advantage of these results. Third, Jacobi iteration is known to many people, so the reader hopefully will not have to puzzle out an unfamiliar algorithm.

Other examples of numerical analysis algorithms, particularly Gaussian elimination have had their performance improved significantly by hand optimizing matrix shapes.[15] This work is based on the CDC Cyber 205. These computers do memory to memory operations, using interleaved memories, so storage patterns which allow data to be accessed sequentially are very advantageous. In the Gaussian elimination example, the shape of the coefficient matrix is studied. The operations in Gaussian elimination are scalar-vector or vector-vector operations, on parts of this matrix. Reshaping the coefficient matrix facilitates fast access to these parts of the matrix. This reshaping is done on a higher level than the individual operations.

Jacobi iteration

Jacobi iteration is one of many methods for solving a system of simultaneous linear equations. The most common methods are direct methods, such as Gaussian elimination, and iterative methods, such as Jacobi iteration, and Gauss-Seidel iteration. The advantage of iterative methods is that they are less computationally complex than direct methods, so much larger systems can be solved in reasonable time. The disadvantage of iterative methods is that, depending on the condition of the matrix, the solution may or may not converge.

The basis for an iterative solution to a system $AX = B$ is to decompose the coefficient matrix A into $M - N$ and attempt to find

X such that $Mx = Nx + B$. A guess X_0 is made, and the algorithm $MX_{n+1} = NX_n + B$ is repeated until X_{n+1} and X_n are close enough in the appropriate norm.

Suppose matrix A is decomposed into 3 parts $A_D + A_L + A_U$. A_D is A with all elements zero except the diagonal. A_L is A with all elements zero except the elements below the diagonal. A_U is A with all elements zero except the elements above the diagonal. The Jacobi iterative scheme is characterized by $M = A_D$ and $N = -(A_L + A_U)$.

This method typifies those methods that do not update variables until the end of a given iteration. It is also well known that for a large class of problems it is the slowest method to converge in terms of the number of iterations necessary for sufficient accuracy.

A one line description of Jacobi iteration appears below.

In the following description, A is the matrix of coefficients of the system. X_i is the solution vector after the i'th iteration. $A_{LU} = -(A_L + A_U)$, the negative of the matrix A with the diagonal zeroed out. A_{DI} is the inverse of the diagonal matrix A_D. B is the right hand side of the system.

In the program, the current solution X_i is multiplied by A_{LU} and B is added. This gives $A_D \times X_{i+1}$. This is multiplied by A_{DI} giving X_{i+1}.

$$X_{i+1} \leftarrow A_{DI} \times (B + A_{LU} \times X_i)$$

Figure 6-1 is a graph of one stage of Jacobi iteration.

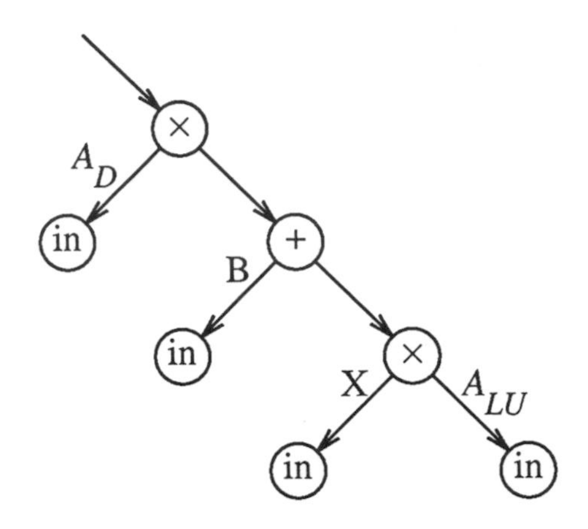

Figure 6-1 Graph of Jacobi Iteration

The effects of shapes on the operations in Jacobi iteration

As we described in the previous section, one stage of Jacobi iteration consists of three operations. In this section we will discuss some of the available methods for doing these operations on a vector machine. These methods involve different shapes for the matrices A_{LU} and A_{DI}.

The first operation is the matrix vector multiplication $A_{LU} \times X_i$.

We will consider three methods for doing this operation.

1. All scalar multiplications and additions.

2. Inner product using vector multiplications and scalar additions.

3. Outer product using vector multiplications and vector additions.

These are based on a machine capable of vector operations. It is assumed that the operations are memory-to-memory and that vector operations are faster if the data items are accessed from contiguous memory locations. The analysis of matrix vector multiplication is due to Leuze and is part of his doctoral dissertation.[19] The following matrix-vector multiplication is analyzed:

$$\begin{bmatrix} a_{11} & a_{12} & a_{13} \\ a_{21} & a_{22} & a_{23} \\ a_{31} & a_{32} & a_{33} \end{bmatrix} \times \begin{bmatrix} x_1 \\ x_2 \\ x_3 \end{bmatrix} = \begin{bmatrix} y_1 \\ y_2 \\ y_3 \end{bmatrix}$$

In the all scalar approach we will use an inner product of each row of A with X_i. No vector instructions are used. Each element of result vector Y is computed as follows.

$$y_1 = a_{11} \times x_1 + a_{12} \times x_2 + a_{13} \times x_3$$

$$y_2 = a_{21} \times x_1 + a_{22} \times x_2 + a_{23} \times x_3$$

$$y_3 = a_{31} \times x_1 + a_{32} \times x_2 + a_{33} \times x_3$$

Calculating the vector Y this way uses n^2 scalar multiplications and $(n-1)^2$ scalar additions.

The second approach is to use inner products with vector multiplication and scalar addition. In this approach we do three vector multiplications. Each row of A is multiplied by X_i.

$$\begin{bmatrix} a_{11} \, a_{12} \, a_{13} \end{bmatrix} \times \begin{bmatrix} x_1 \\ x_2 \\ x_3 \end{bmatrix}$$

$$\begin{bmatrix} a_{21} \, a_{22} \, a_{23} \end{bmatrix} \times \begin{bmatrix} x_1 \\ x_2 \\ x_3 \end{bmatrix}$$

$$\begin{bmatrix} a_{31} \, a_{32} \, a_{33} \end{bmatrix} \times \begin{bmatrix} x_1 \\ x_2 \\ x_3 \end{bmatrix}$$

The elements y_1, y_2, and y_3 are the sum of the first, second, and third result vectors.

$$y_1 = a_{11} x_1 + a_{12} x_2 + a_{13} x_3$$

$$y_2 = a_{21} x_1 + a_{22} x_2 + a_{23} x_3$$

$$y_3 = a_{31} x_1 + a_{32} x_2 + a_{33} x_3$$

If the multiplications are done by vector operations and the additions are scalar, the time for the method is n vector multiplications of length n and $(n-1)^2$ scalar additions. For best efficiency this method requires that the matrix A be in row order.

The third method we will describe is an outer product method using both vector multiplications and vector additions. In this approach we multiply each column of A with an element of X_i.

$$\begin{bmatrix} a_{11} \\ a_{21} \\ a_{31} \end{bmatrix} \times x_1 = \begin{bmatrix} a_{11}x_1 \\ a_{21}x_1 \\ a_{31}x_1 \end{bmatrix}$$

$$\begin{bmatrix} a_{12} \\ a_{22} \\ a_{32} \end{bmatrix} \times x_2 = \begin{bmatrix} a_{12}x_2 \\ a_{22}x_2 \\ a_{32}x_2 \end{bmatrix}$$

$$\begin{bmatrix} a_{13} \\ a_{23} \\ a_{33} \end{bmatrix} \times x_3 = \begin{bmatrix} a_{13}x_3 \\ a_{23}x_3 \\ a_{33}x_3 \end{bmatrix}$$

The result Y is determined by summing the three vectors with vector addition operations.

$$\begin{bmatrix} a_{11}x_1 \\ a_{21}x_1 \\ a_{31}x_1 \end{bmatrix} + \begin{bmatrix} a_{12}x_2 \\ a_{22}x_2 \\ a_{32}x_2 \end{bmatrix} + \begin{bmatrix} a_{13}x_3 \\ a_{23}x_3 \\ a_{33}x_3 \end{bmatrix} = \begin{bmatrix} y_1 \\ y_2 \\ y_3 \end{bmatrix}$$

This method requires that the matrix A be in column major order for greatest efficiency.

This method requires n vector multiplications of length n and $n-1$ vector additions of length n.

The result of multiplying A_{LU} by x_i is the temporary vector Y.

The second step in Jacobi iteration is to add the right hand side B to this. Assuming that only one shape is available to vectors, there is only one option for doing this operation: a single vector addition of length n. The temporary vector which is the result of this addition will be called z.

The final step in Jacobi iteration is to multiply the result of

$$B + A_{LU} \times x_i$$

or z by A_{DI}. This is also a matrix vector multiplication so the three choices given for the first step also apply. However, A_{DI} is a diagonal matrix. Multiplying a vector z by A_{DI} is equivalent to multiplying each element z_i by the corresponding element $1/a_{ii}$ on the diagonal. A_{DI} could be stored in an additional shape, as a vector consisting of the elements on the diagonal. The multiplication $A_{DI} \times z$ could be done as a single vector multiply of length n.

$$\begin{bmatrix} 1/a_{11} \\ 1/a_{22} \\ 1/a_{33} \end{bmatrix} \times \begin{bmatrix} z_1 \\ z_2 \\ z_3 \end{bmatrix}$$

Cost functions for the three steps

In the next sections cost functions are given for the steps of Jacobi iteration. These functions consist of tables which are indexed by the shapes of the matrices involved in the operations and the different methods which do the operations. The vector operands are only allowed to have one shape. Since a vector is stored sequentially this "shape" will be called "sequential".

The cost tables are based on the methods described above. The matrix A is assumed to be 100 by 100. The table of instruction timings for scalar and vector additions and multiplications are from the CDC Cyber 205.

The timings are in 20 nanosecond cycles. S is the startup overhead for a vector pipeline. The length of a vector is called n. L is the number of results per cycle which emerge from the pipe. The time for a vector instruction of length n is $S + n/L$. The column labeled "scalar" gives the time, in 20 nanosecond cycles, for one corresponding scalar operation, producing one result.

instruction	S	L	scalar
add	51	4	5
subtract	51	4	5
multiply	52	4	5
divide	80	.56	50

Table 6-1: timing for vector instructions

The following are the cost tables for the three operations of Jacobi iteration. These tables can be used to make the minimum cost shape assignment to each of the data structures in the program segment.

method	res shape	left op shape	rt op shape	cost
all scalar	seq	any shape	sequential	100000
inner prod	seq	row order	sequential	57700
outer prod	seq	column order	sequential	15300

Table 6-2: cost table for matrix vector multiplication: full matrix

method	res shape	left op shape	rt op shape	cost
scalar add	seq	sequential	sequential	500
vector add	seq	sequential	sequential	76

Table 6-3: cost table for vector addition

method	res shape	left op shape	rt op shape	cost
all scalar	seq	any shape	sequential	100000
inner prod	seq	row order	sequential	57700
outer prod	seq	column order	sequential	15300
vector mult	seq	diag as vect	sequential	77

Table 6-4: Cost table for matrix vector multiply, diagonal matrix

Using the tree algorithm to solve the shapes problem on a graph of Jacobi iteration

This section is a description of applying the tree algorithm to the Jacobi iteration graph. For convenience, each node is labeled with a letter. Each arc is labeled with the same letter as the node for which it is the parent.

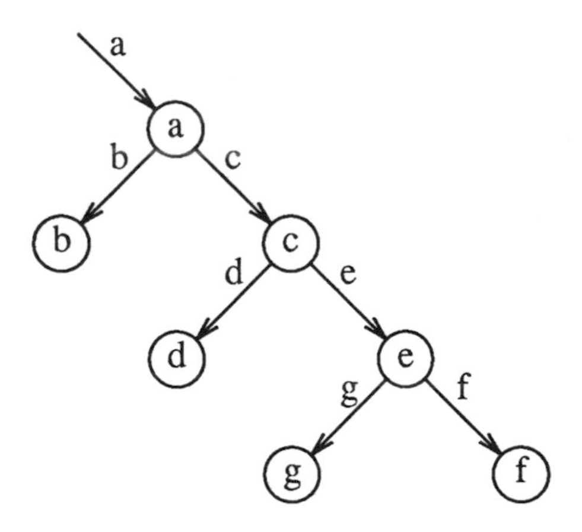

Figure 6-2 Jacobi iteration with arcs labeled with letters

The following is a list, for each arc α, of the set of shapes S_α

which may be assigned to that arc. This depends on whether the data item which flows along the arc is a full matrix, a diagonal matrix, or a vector. Vectors are only allowed one "shape", sequential storage.

$$
\begin{aligned}
S_a &= \{\text{sequential}\} \\
S_b &= \{\text{row order, column order, diagonal as vector}\} \\
S_c &= \{\text{sequential}\} \\
S_d &= \{\text{sequential}\} \\
S_e &= \{\text{sequential}\} \\
S_f &= \{\text{row order, column order}\} \\
S_g &= \{\text{sequential}\}
\end{aligned}
$$

Because node g represents x_i and node a represents x_{i+1}, the shapes on arc a and arc g are forced to be the same. Since this data item is a vector, and a vector only has one shape, this is always the case.

The shapes problem can be solved on this graph by applying the tree algorithm three times: to node e (Table 6-2), node c (Table 6-3), and node a (Table 6-4).

1) There are three methods which can be used to compute node e: all scalar, inner product with vector multiplication and scalar addition, and outer product with vector multiplication and vector addition (see table 6-2). Since the result is a vector, there is only one possible shape for the result of node e. So, step 1 of the minimization finds a single minimum cost for computing node e into a vector. This amounts to finding the minimum entry in table 6-2. Step 2, the transpose, is nonexistent, since there is only one

possible shape for the result. In the cost function minimization below, each method represents one line of table 6-2. M_1 is the all scalar method, M_2 is the vector inner product, and M_3 is the vector outer product. Since nodes f and g are leaf nodes, the cost of getting each of them into any shape is zero.

$$
\begin{aligned}
cost(e, S_e) = \min \{ & (M_1.opcost + \\
& \quad cost(f, any\ shape) + \\
& \quad cost(g, any\ shape)), \\
& (M_2.opcost + \\
& \quad cost(f, row) + \\
& \quad cost(g, sequential)), \\
& (M_3.opcost + \\
& \quad cost(f, column) + \\
& \quad cost(g, sequential))\}
\end{aligned}
$$

$$
\begin{aligned}
cost(e, S_e) = \min \{ & (10000 + 0 + 0), \\
& (57700 + 0 + 0), \\
& (15300 + 0 + 0)\}
\end{aligned}
$$

$$cost(e, S_e) = 15300$$

The least cost method is the vector outer product method. This assigns the column order shape to arc f. Arc g would be assigned the sequential shape.

2) There are two methods which can compute node c: scalar addition and vector addition (see table 6-3). The minimization here simply amounts to choosing the single minimum cost entry in table 6-3. Again, the result can only be in 1 shape, so the transpose step is nonexistent. In this cost function minimization, M_1

represents scalar addition and M_2 represents vector addition.

$$
\begin{aligned}
cost(c, S_c) = \min \{ & (M_1.opcost + \\
& \quad cost(d, sequential) + \\
& \quad cost(e, sequential)), \\
& (M_2.opcost + \\
& \quad cost(d, sequential) + \\
& \quad cost(e, sequential))\}
\end{aligned}
$$

$$
\begin{aligned}
cost(c, S_c) = \min \{ & (500 + 0 + 15300), \\
& (76 + 0 + 15300)\}
\end{aligned}
$$

$$cost(c, S_c) = 15376$$

The least cost method is the vector addition. Arcs d and e are assigned the sequential shape.

3) There are four methods which can be used to compute node a: all scalar, inner product with vector multiplication and scalar addition, outer product with vector multiplication and addition, and vector multiplication by storing the diagonal as a vector (see table 6-4). Again, the result can only be in 1 shape, so the minimization consists of simply choosing the single minimum entry from table 6-4. Again, the transpose step is trivial. In this minimization, M_1 represents all scalar multiplication, M_2 represents vector inner product, M_3 represents vector outer product, and M_4 represents vector multiplication of the diagonal.

$$
\begin{aligned}
cost(a, S_a) = \min \{ &(M_1.opcost + \\
&\quad cost(b, any\ shape) + \\
&\quad cost(c, sequential)), \\
&(M_2.opcost + \\
&\quad cost(b, row\ order) + \\
&\quad cost(c, sequential)) \\
&(M_3.opcost + \\
&\quad cost(b, column\ order) + \\
&\quad cost(c, sequential)), \\
&(M_4.opcost + \\
&\quad cost(b, diag\ as\ vector) + \\
&\quad cost(c, sequential])\}
\end{aligned}
$$

$$
\begin{aligned}
cost(a, S_a) = \min \{ &(10000 + 0 + 15376), \\
&(57000 + 0 + 15376), \\
&(15300 + 0 + 15376), \\
&(77 + 0 + 15376)\}
\end{aligned}
$$

$$
cost(a, S_a) = 15453
$$

The tree algorithm would choose the vector multiplication of the diagonal, and the diagonal matrix which corresponds to arc b would be stored as a single vector consisting of the diagonal of the matrix.

The preceding application of the tree algorithm was very straightforward. This was particularly so because the two matrices were both leaf nodes, so the cost of obtaining them in all possible shapes was zero.

Despite the simplicity of the process, the results are very

rewarding. The cost for the shape assignment found by the tree algorithm is 15,453 cycles. An arbitrary assignment of row order to both matrices would give a cost of 115,476 cycles.

Solving the shapes problem on the Jacobi iteration graph using the collapsible graph algorithm

The shapes problem can also be solved, achieving the same result of course, by six applications of transformation A. Each time transformation A is applied, a shape is assigned to an arc, and that arc is deleted from the graph.

In order to do this, cost tables are specified for all the nodes in the graph, including the leaf nodes. The cost tables for nodes which are operator nodes are the same as in the tree algorithm example. An index for the shape on the arc representing the result is included in these tables.

method	S_a	S_b	S_c	T_a
scalar	seq	row	seq	100000
scalar	seq	col	seq	100000
inner prod	seq	row	seq	57700
outer prod	seq	col	seq	15300
vector mult	seq	diag as vector	seq	77

Table 6-5: cost of computing operand $a = b \times c$

S_b	$T_b[S_b]$
row	0
col	0
diag/vector	0

Table 6-6: cost of reading operand b

method	S_c	S_d	S_e	T_c
scalar add	seq	seq	seq	500
vector add	seq	seq	seq	76

Table 6-7: cost of computing result $c = d + e$

S_d	$T_D[S_d]$
seq	0

Table 6-8: cost of reading operand d

method	S_e	S_f	S_g	T_e
scalar mult	seq	row	seq	100000
scalar mult	seq	col	seq	100000
inner prod	seq	row	seq	57700
outer prod	seq	col	seq	15300

Table 6-9: cost of computing operand $e = f \times g$

S_f	$T_f[S_f]$
row	0
col	0

Table 6-10: cost of reading operand f

S_g	$T_g[S_g]$
seq	0

Table 6-11: cost of reading operand g

In the following six steps, transformation A is applied to collapse the graph. After the graph is collapsed, an optimal cost shape assignment can be made.

1) Nodes e and f are merged by transformation A to form node ef. This step combines the cost function for node e, showing in Table 6-9 with the cost function for node f shown in Table 6-10.

$$T_{ef}[S_e, S_g] = \min \{(T_f[row] + T_e[S_e, row, S_g]), (T_f[column] + T_e[S_e, column, S_g])\}$$

The resulting table T_{ef} is shown below:

S_e	S_g	$T_{ef}[S_e, S_g]$
seq	seq	15300 (S_f = col)

Table 6-12: cost of newly merged node ef

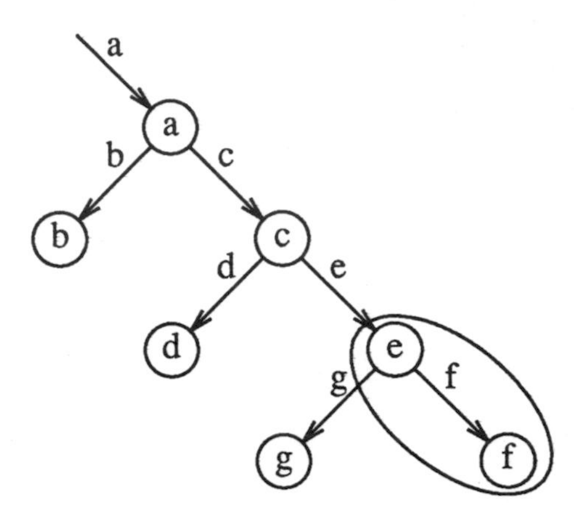

Figure 6-3 Step 1: nodes E and F are merged by Transformation A

2) Node ef and g are merged to form efg. In this step we combine the cost function for node g, shown in Table 6-11, with the cost function for node ef, shown in Table 6-12.

$$T_{efg}[S_e] = \min \{(T_{ef}[S_e, sequential] + T_g[sequential])\}$$

Since there is only one possible shape for S_g, the minimization only has one choice. For convenience, node efg is renamed node e. The resulting table is:

S_e	$T_e[S_e]$
seq	15300 (S_g = sequential)

Table 6-13: cost of new node e (renamed from merged node efg)

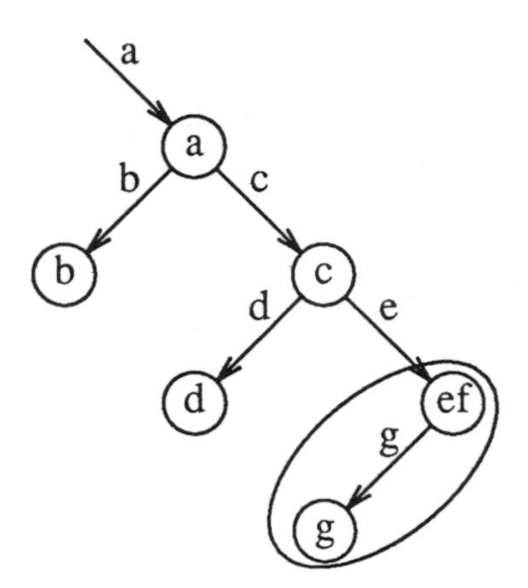

Figure 6-4 Step 2: nodes EF and G are merged by Transformation A

3) Nodes c and d are merged to form node cd. In this step we combine the cost function for node d, shown in Table 6-8, with the cost function for node e, shown in Table 6-13.

$$T_{cd}[S_c, S_e] = \min \{(T_c[S_c, sequential, S_e] + T_d[sequential])\}$$

Since S_d can only have one shape, this minimization has only one choice. The resulting table is:

S_c	S_e	$T_{cd}[S_c, S_e]$
seq	seq	76 (S_d = sequential)

Table 6-14: cost of new node cd

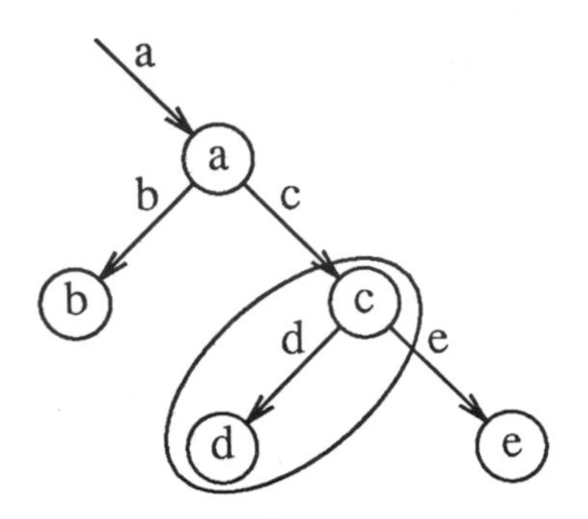

Figure 6-5 Step 3: nodes c and d are merged by Transformation A

4) Nodes cd and e are merged to form node cde.

$$T_{cde}[S_c] = \min\{(T_{cd}[S_c, sequential] + T_e[sequential])\}$$

This is another example in which the arc can only have one shape, so the minimization has one choice. For convenience, node cde is renamed node c. The table is:

S_c	$T_c[S_c]$
seq	15376 (S_e = sequential)

Table 6-15: cost of new node c (renamed from cde)

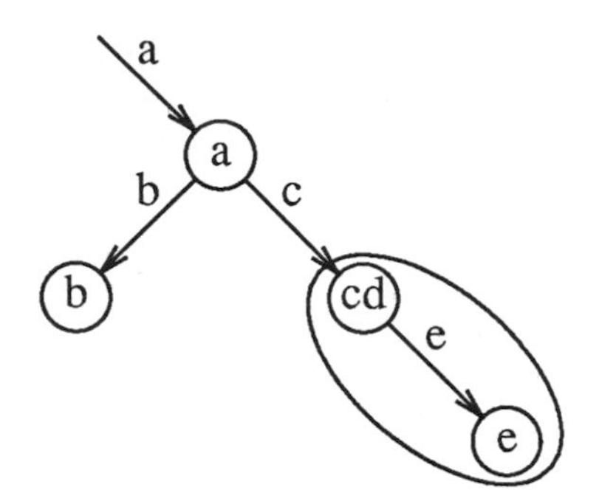

Figure 6-6 Step 4: nodes cd and e are merged by transformation A

5) Nodes a and b are combined to form node ab.

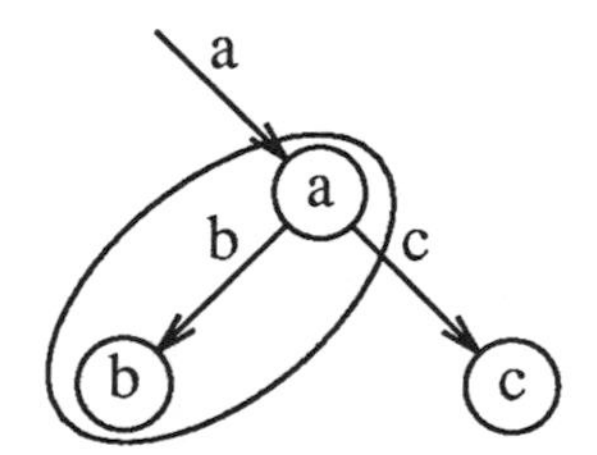

Figure 6-7 Step 5: nodes a and b are merged by transformation A

$$T_{ab}[S_a, S_c] = \min \{$$
$$(T_a[S_a, row, S_c] + T_b[row]),$$
$$(T_a[S_a, column, S_c] + T_b[column]),$$
$$(T_a[S_a, diag\ as\ vector, S_c] + T_b[diag\ as\ vector])\}$$

The resulting table is:

S_a	S_c	$T_{ab}[S_a, S_b]$
seq	seq	77 (S_b = diag as vector)

Table 6-16: cost of new node ab

6) Nodes ab and c are combined to form node abc.

$$T_{abc}[S_a] = \min\{(T_{ab}[S_a, S_c] + T_c[S_c])\}$$

This is another case where the arc can have only one shape so the minimization has only one choice. The resulting table is:

S_a	$T_{abc}[S_a]$
seq	15453 (S_c = sequential)

Table 6-17: cost of new node abc

At the end of the process the following optimal shape assignment has been made: S_b = diag as vector, S_c = sequential, S_d = sequential, S_e = sequential, S_f = column order, S_g = sequential. The same cost and shape assignment that the tree algorithm found was also found by using transformation A of the collapsible graph algorithm.

Appendix A

DEFINITION OF COLLAPSIBLE GRAPHS

In the following discussion, i, j, and k are nodes. The indegree of a node i, which will be called *indeg* (i), is the number of incoming arcs. The outdegree of a node i, or *outdeg* (i) is the number of outgoing arcs. The notation $t(i, j)$ where t is one of $\{a, b, c\}$ indicates that t is applied to nodes i and j. When no specific nodes are being referenced, the (i, j) part of the notation will be left out.

If a transformation $t(i, j)$ merges nodes i and j, a new node ij is created. The arcs which touched i and j before $t(i, j)$ now touch ij, except for any arcs joining i and j, which are deleted.

Definition: A directed acyclic graph G is collapsible if it can be reduced to a single node by a sequence of the following operations:

Transformation $a(i, j)$ can be applied to a subgraph with two nodes i and j such that i is a parent of j, and one of i and j has indeg ≤ 1 and outdeg $= 1$. Transformation $a(i, j)$ merges nodes i and j and deletes arc (i, j).

Case 1: i (the parent node) has indeg $= 1$ and outdeg $= 1$. After $a(i, j)$ the new node ij has: $indeg(ij) \leq indeg(j)$, $outdeg(ij) = outdeg(j)$.

Case 2: j (the child node) has indeg $= 1$ and outdeg $= 1$. After $a(i, j)$ the new node ij has: $indeg(ij) = indeg(i)$, $outdeg(ij) = outdeg(i)$.

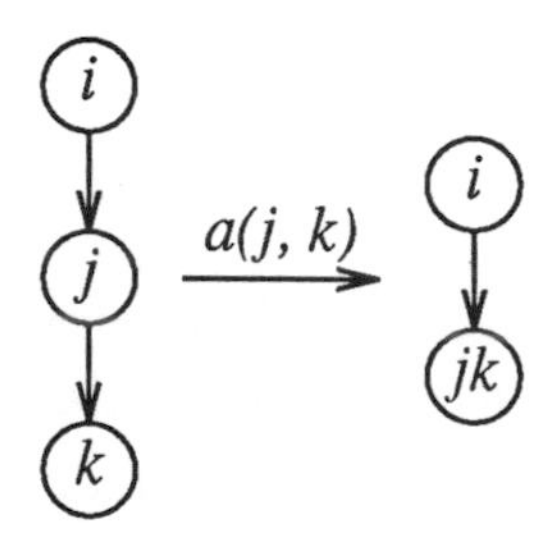

Figure 7-1 Transformation a

Transformation $b(i, j)$ transforms a subgraph consisting of two nodes i and j such that i is a parent of j, and there are two or more arcs from i to j, into a subgraph consisting of two nodes i and j connected by a single arc. $Indeg(i)$ and $outdeg(j)$ are unchanged.

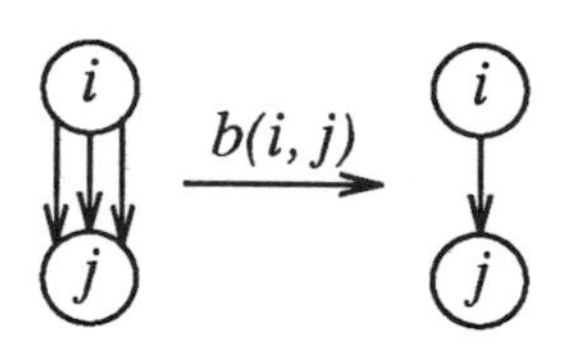

Figure 7-2 Transformation b

Transformation $c(i, j)$ transforms a subgraph consisting of two nodes i and j such that i is a parent of j, and $indeg(j) = 1$ and $outdeg(j) = 0$, into a subgraph consisting of a single node ij such that $indeg(ij) = indeg(i)$ and $outdeg(ij) = outdeg(i) - 1$.

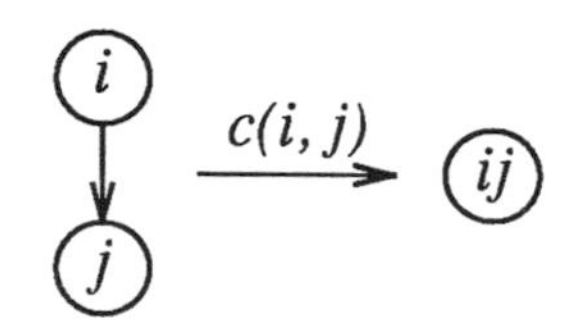

Figure 7-3 Transformation C

This class of graphs was first defined by Prabhala and Sethi.[21]

Appendix B
RESTRICTION 1

The restriction that each node has indegree ≤ 1 or outdegree ≤ 1 will be imposed on the collapsible graphs. This restriction will be referred to as *Restriction 1*. Restriction 1 is required for the time bound on this algorithm.

Any collapsible graph G that violates Restriction 1 and that represents a program can be changed to a graph G' that satisfies Restriction 1 as follows: Suppose node *op* has indegree = i and outdegree = o, i and $o > 1$. Node *op* is divided into 2 nodes *tr* and *op*. *Tr* has indegree = i and outdegree = 1. The single arc which leaves *tr* goes to *op*. *Op* has indegree = 1 and outdegree = o. If all nodes in G which have indegree > 1 and outdegree > 1 are divided in this way, then a new graph G' is formed which satisfies Restriction 1.

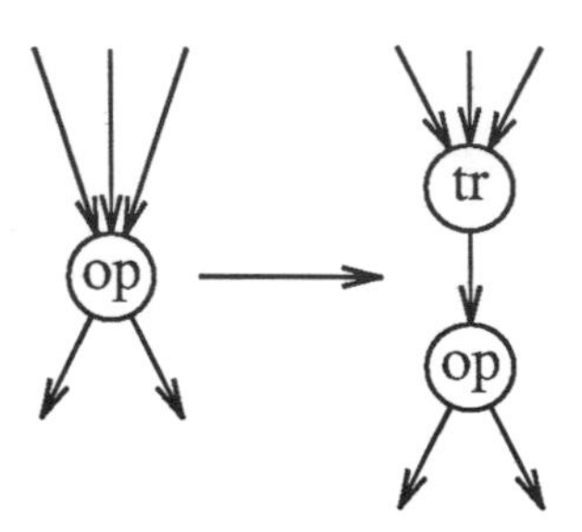

Figure 7-4 Restriction 1

When a node is conveniently divided to satisfy Restriction 1, the question arises: what happens to the cost function? Since the node costs are added to form the total cost for the graph, the cost function of a divided node must be divided additively. The following is an argument for the natural division into two additive parts of the cost function of a node with indegree > 1 and outdegree > 1. When a node with indegree > 1 and outdegree > 1 is divided into two parts the cost function is handled as follows.

Assume that there are no methods for function computation that compute the result in more than one shape simultaneously. Since the interpretation of the graph is a program, the node with indegree > 1 and outdegree > 1 is a node which is computed and shared. A node with these characteristics would have a two part cost function. One part is the cost of computing the result, which depends on the shapes of the operands and the result. The other part is the cost of transposing the result into the lowest cost set of shapes for the operations which share the result.

When a node is divided to make the indegree ≤ 1 or outdegree

≤ 1 requirement true, the part of the cost associated with the computation is associated with the lower node, *op*. The arcs which touch this node correspond to the operands and a single computed result. The part of the cost function which represents the transpositions is associated with the upper of the two new nodes, *tr*. The arcs which touch this node correspond to the computed result, and all the shared node arcs. The total cost of the original composite node can thus always be expressed as the sum of these two cost components.

Appendix C

PROPERTIES OF COLLAPSIBLE GRAPH TRANSFORMATIONS

Theorem 3.1: If Restriction 1 is true for graph G then it is true for $t(G)$, where t is one of $\{A, B\}$.

Proof: A and B do not change indegree or outdegree of any node outside the subgraph. $A(i, j)$ and $B(i, j)$ both merge nodes i and j into one node ij. All that is needed is to show that node ij meets Restriction 1.

Transformation A preserves Restriction 1:

$Indeg(ij) = indeg(i)$ and $outdeg(ij) = outdeg(i) - t$ in $A(i, j)$, so indeg and outdeg of ij are equal to or less than indegree and outdegree of node i. So if node i meets

Restriction 1, so does node ij.

Transformation B preserves Restriction 1:

After transformation $B(i, j)$ is applied: $indeg(ij) = indeg(i) + indeg(j) - t$ and $outdeg(ij) = outdeg(i) + outdeg(j) - t$.

Case 1: $outdeg(i) = t$ and $outdeg(j) \leq 1$ so $outdeg(ij) \leq 1$.

Case 2: $indeg(j) = t$ and $indeg(i) \leq 1$ so $indeg(ij) \leq 1$.

Case 3: $indeg(i) = 1$ and $indeg(j) = 1$ and $t = 1$ so $indeg(ij) = 1$.

Theorem 3.2: If Restriction 1 is true for graph G then it is true for $t(G)$ where t is one of $\{a, b, c\}$.

$a(i, j)$ preserves Restriction 1:

Case 1: i has $indeg \leq 1$ and $outdeg$=1: After $a(i, j)$, $indeg(ij) \leq indeg(j)$ and $outdeg(ij) = outdeg(j)$. So if Restriction 1 is true for node j, it will be true for node ij.

Case 2: j has indeg = 1 and outdeg = 1: After $a(i, j)$, $indeg(ij) = indeg(i)$ and $outdeg(ij) = outdeg(i)$. So if Restriction 1 is true for node i, it will be true for node ij.

$b(i, j)$ preserves Restriction 1:

$b(i, j)$ does not affect $indeg(i)$ or $outdeg(j)$. $b(i, j)$ decreases $outdeg(i)$ and $indeg(j)$. So Restriction 1 is preserved.

$c(i, j)$ preserves Restriction 1:

After doing $c(i, j)$, $indeg(ij) = indeg(i)$ and $outdeg(ij) = outdeg(i) - 1$ so Restriction 1 is preserved.

Theorem 3.3: A and B do not cause the degree of the graph to grow.

Proof: After $A(i, j)$, $indeg(ij) = indeg(i)$ and $outdeg(ij) = outdeg(i) - t$ so the total degree of ij is less than the total degree of i.

After $B(i, j)$, $indeg(ij) = indeg(j) + indeg(i) - t$ and $outdeg(ij) = outdeg(j) + outdeg(i) - t$.

Case 1 of B: $Indeg(i) \leq 1$ and $t \geq 1$ so $indeg(ij) \leq indeg(j)$. $Outdeg(i) = t$ so $outdeg(ij) = outdeg(j)$. Since $indeg(ij) \leq indeg(j)$ and $outdeg(ij) = outdeg(j)$ the total degree of ij is less than or equal to the total degree of j.

Case 2 of B: $Indeg(j) = t$ so $indeg(ij) = indeg(i)$. $Outdeg(j) = 1$ and $t \geq 1$ so $outdeg(ij) \leq outdeg(i)$. Since $indeg(ij) = indeg(i)$ and $outdeg(ij) \leq outdeg(i)$ the total degree of ij is less than or equal to the total degree of i.

Case 3 of B: $Indeg(i) = 1$ and $t = 1$ so $indeg(ij) = indeg(j)$. $Outdeg(i) = 1$ and $t = 1$ so $outdeg(ij) = outdeg(j)$. Since $indeg(ij) = indeg(j)$ and $outdeg(ij) = outdeg(j)$ the total degree of ij equals the total degree of j.

Appendix D

EQUIVALENCE OF {a, b, c} TO {A, B}

This section demonstrates that a sequence S of transformations $\{a, b, c\}$ that collapses a graph G can be changed in several steps to a sequence V of transformations $\{A, B\}$ that also collapses G.

Since one of the steps in this process involves permuting the order of $\{a, b, c\}$ we first show in Lemma 3.1 that applying any one of $\{a, b, c\}$ that is allowable leaves the graph collapsible.

The first group of lemmas, 3.2 through 3.6, describes the relationships between transformations $\{a, b, c\}$ and $\{A, B\}$; and when A and B can be substituted for a, b, and c.

Lemma 3.2 shows that $c(i, j)$ can be replaced by $A(i, j)$, and

Lemma 3.3 shows that $a(i, j)$ can be replaced by $B(i, j)$.

Lemma 3.4 shows that the two cases of transformation a produce subgraphs which are isomorphic to each other, so if both cases were applicable, one could be substituted for the other in a sequence, while still leaving the graph collapsible.

Lemma 3.5 shows that the consecutive pair of transformations $b(i, j)a(i, j)$ can be replaced by $B(i, j)$, and lemma 3.6 shows that the consecutive pair of transformations $b(i, j)c(i, j)$ can be replaced by $A(i, j)$.

The second group of lemmas, 3.7 through 3.12, are concerned with transforming the sequence S into the sequence V. The idea is to show that every b in S is associated with an a or c which deletes the arc that the b introduced. The b is then moved forward in the sequence until it is just before its associated a or c. The first step in this process is to establish a one to one relationship of b's with a's or c's.

Lemmas 3.7, 3.8, and 3.9 show that S can be changed to an intermediate sequence S'. S' only includes occurrences of $b(i, j)$ such that there are no multiple arc paths from i to j. This prevents a sequence in which $b(i, j)$ is applied, then the paths from i to j are collapsed to arcs, and $b(i, j)$ must be applied again.

Lemma 3.10 then states that for every b in S', there exists an a or c later in S' that deletes the arc introduced by b.

Lemma 3.11 shows that S' can be changed to a second intermediate sequence S''. In S'' the first a applicable to a subgraph with nodes i, j, k where $indeg(j) = 1$ and $outdeg(j) = 1$ applies to one of the arcs (i, j) or (j, k) which was introduced by a b, if either arc was so introduced. The isomorphism result of lemma

3.4 is used in creating S''.

Lemma 3.12 shows that S'' can be changed to a third intermediate sequence S''' where each $b(i,j)$ is moved forward in the sequence until it is just before the $a(i,j)$ or $c(i,j)$ that is associated with it. This lemma shows that for any transformation t which is one of $\{a,b,c\}$ before $a(i,j)$ or $c(i,j)$, if $b(i,j)$ is moved from before t to after t, t still applies.

Finally, Theorem 3.4 substitutes the A or B transformations for the a,b,c. First, every pair $b(i,j)a(i,j)$ is replaced by $B(i,j)$ and every pair $b(i,j)c(i,j)$ is replaced by $A(i,j)$. This results in a fourth intermediate sequence S'''' which contains no occurrence of b. Every remaining a in S'''' is replaced by B and every remaining c in S'''' is replaced by A. The resulting sequence V contains only A and B.

Lemma 3.1: If G is a collapsible graph and G can be transformed to G' by one of transformations $\{a,b,c\}$ then G' is a collapsible graph.

Proof: A proof of this property can be found in Prabhala and Sethi.[21]

We can therefore carry out any applicable transformation at any point without hurting our chances of ultimately reducing the graph to a single node.[17]

Lemma 3.2: An $a(i,j)$ can be replaced by case 3 of transformation $B(i,j)$.

Proof: case 3 of $B(i,j)$ can be applied to a subgraph in which a node i has indeg ≤ 1 and outdeg $= 1$. These requirements are the same as the requirements for transformation $a(i,j)$. Case 3 of transformation $B(i,j)$ merges nodes i and j and deletes the arc between them. The resulting node ij has: $indeg\,(ij) = indeg\,(j)$, and $outdeg\,(ij) = outdeg\,(j)$, which is the same as the merged node created by $a\,(i\,,j)$.

Lemma 3.3: A $c\,(i\,,j)$ can be replaced by transformation $A\,(i\,,j)$.

Proof: If t, the number of arcs connecting i and j is 1, transformation $A\,(i\,,j)$ applies to the same subgraph as $c\,(i\,,j)$. The resulting subgraph, a single node ij such that $indeg\,(ij) = indeg\,(i)$, $outdeg\,(ij) = outdeg\,(i) - 1$, is the same for both $A\,(i\,,j)$ and $c\,(i\,,j)$.

Lemma 3.4: Given a subgraph consisting of three nodes i, j, and k such that i is the parent of j, j is the parent of k, $indeg\,(j) = outdeg\,(j) = 1$, transformation a can be applied in one of two ways. These are case 1: $a\,(j\,,k)$, and case 2: $a\,(i\,,j)$. The subgraphs resulting from case 1 and case 2 of transformation a are isomorphic to each other.

Proof: If $a\,(j\,,k)$ is applied, the new subgraph consists of two nodes i and jk, connected by a single arc. $Indeg\,(i)$ and $outdeg\,(i)$ are unchanged and $indeg\,(jk) = indeg\,(k)$, $outdeg\,(jk) = outdeg\,(k)$. If $a\,(i\,,j)$ is applied, the new subgraph consists of two nodes ij and k, connected by a single arc. $Indeg\,(ij) = indeg\,(i)$, $outdeg\,(ij) = outdeg\,(i)$. $Indeg\,(k)$ and $outdeg\,(k)$ are unchanged. The two subgraphs are identical except for the labeling.

Lemma 3.5: The consecutive pair of transformations $b\,(i\,,j)a\,(i\,,j)$ can be replaced by $B\,(i\,,j)$ whenever the original sequence was legal.

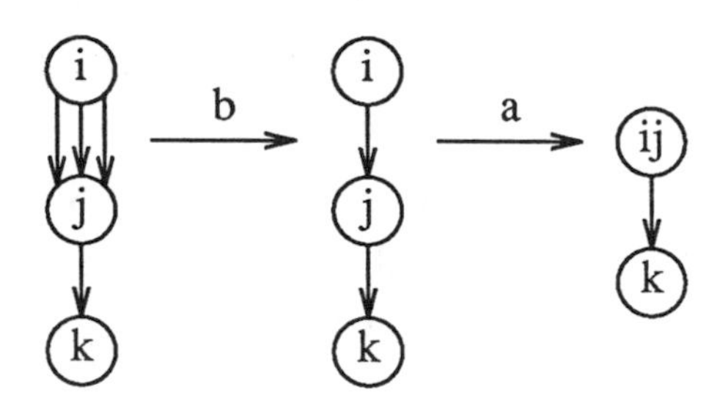

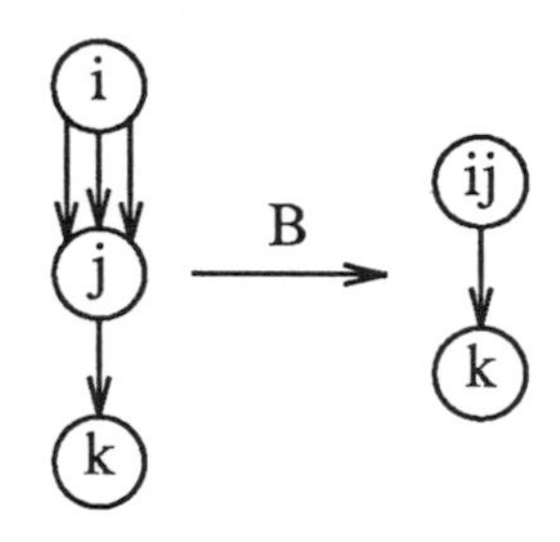

Figure 7-5 Equivalence of sequence {ba} to transformation B

Proof: we will show that if the requirements for $b(i,j)a(i,j)$ are met by a subgraph, then $B(i,j)$ can be applied with the same results. Given a subgraph with two nodes i and j, in order to apply $b(i,j)$ and then immediately apply $a(i,j)$, several restrictions must apply to the subgraph. There must be more than one arc between i and j. Because of Restriction 1, node i is forced to have indeg ≤ 1.

(1) Node j is the only child of i: in this case i has indeg = 1 and outdeg = 1 after $b(i,j)$, so case 1 of $a(i,j)$ can be applied. Case 1 of $B(i,j)$ only requires that $indeg(i) \leq 1$ and that node j is the only child of i.

(2) Node i is the only parent of j: in this case j has indeg = 1

and outdeg = 1 after $b(i, j)$, so case 2 of $a(i, j)$ can be applied. Case 2 of $B(i, j)$ only requires that $indeg(i) \leq 1$ and that node i is the only parent of j.

(3) Neither (1) or (2) above is the case. Since (1) does not hold, i has a child j' other than j, and since (2) does not hold, j has a parent i' other than i. After $b(i, j)$, $indeg(j) \geq 2$, since both i and i' are j's parents. Similarly, $outdeg(i) \geq 2$ since both j and j' are i's children. Since neither j nor i has indeg ≤ 1 and outdeg ≤ 1, $a(i, j)$ cannot be applied, and thus the original $b(i, j)a(i, j)$ sequence was not legally applicable.

$B(i, j)$ produces the same subgraph as $b(i, j)a(i, j)$. The $b(i, j)$ replaces t arcs between i and j by a single arc (i, j). The $a(i, j)$ deletes arc (i, j) and merges nodes i and j. The resulting subgraph consists of a node ii such that $indeg(ii) = indeg(i) + indeg(i) - t$, and $outdeg(ij) = outdeg(i) + outdeg(j) - t$. The $B(i, j)$ removes the t arcs joining i and j and merges i and j. The resulting subgraph consists of a node ij with the following indegree and outdegree: $indeg(ij) = indeg(j) + indeg(i) - t$, $outdeg(ij) = outdeg(j) + outdeg(i) - t$.

Lemma 3.6: The consecutive pair of transformations $b(i, j)c(i, j)$ can be replaced by $A(i, j)$ whenever the original sequence was legal.

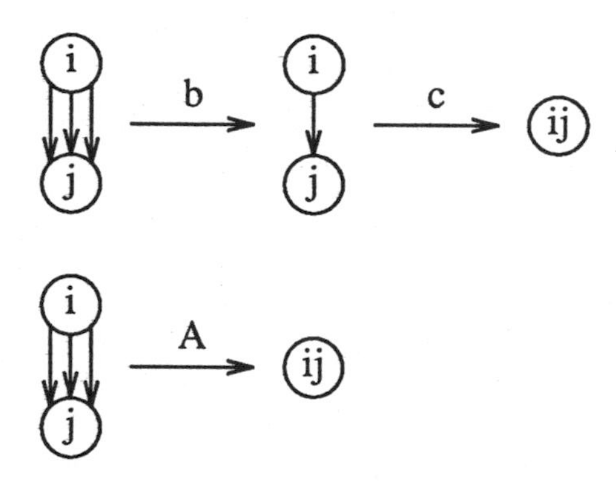

Figure 7-6 Equivalence of sequence {bc} to transformation A

Proof: We will show that if the requirements for $b(i,j)c(i,j)$ are met by a subgraph, then $A(i,j)$ can be applied with the same results. Given a subgraph with two nodes i and j, in order to apply $b(i,j)$ immediately followed by $c(i,j)$ certain restrictions must apply to the subgraph. First, there must be several arcs joining i and j. Second, $outdeg(j) = 0$. Third, i must be the only parent of j. This last requirement is necessary so that after $b(i,j)$ replaces all the arcs joining i and j by a single arc no other arcs will touch j and $c(i,j)$ can be applied immediately.

The requirements for $A(i,j)$ are just that i is the only parent of j, and that $outdeg(j) = 0$, so if the requirements for $b(i,j)c(i,j)$ are met, $A(i,j)$ can also be applied.

The result of applying $b(i,j)c(i,j)$ is a subgraph consisting of a single node ij such that $indeg(ij) = indeg(i)$ and $outdeg(ij) = outdeg(i) - t$. Applying $A(i,j)$ deletes all t arcs between i and j and merges the two nodes. The resulting node ij has $indeg(ij) = indeg(i)$ and $outdeg(ij) = outdeg(i) - t$. So the subgraph resulting from $A(i,j)$ is identical to the subgraph resulting from

$b(i, j)c(i, j)$.

Lemma 3.7: Given a subgraph with two nodes j and k such that j and k are connected by one or more paths greater than one arc in length, the paths from j to k will be collapsed by the remainder of S to one or more single arcs from j' to k', where j may have been merged with nodes $j_1 \cdots j_n$ to form j' and k may have been merged with nodes $k_1 \cdots k_m$ to form k'.

Proof: Since S collapses G to a single node, at some point in S, nodes j and k must be merged. Before j is merged with k, j may be merged with nodes $j_1 \cdots j_n$, and k may be merged with nodes $k_1 \cdots k_m$. The only transformations that can merge nodes j' and k' are a and c. In order for $a(j', k')$ and $c(j', k')$ to be applied, j' and k' must be joined by a single arc (j', k'). If arc (j', k') was introduced by a b, then before the b, there was more than one arc from j' to k'. Therefore, at some point in S, j' and k' were joined by one or more arcs.

Lemma 3.8: Given two nodes j and k such that j and k are connected by one or more paths of length greater than one arc, and j and k are also connected by t arcs, where $t \geq 1$: any transformation that can be applied to j or k if $t = 1$ can be applied to j or k if $t > 1$.

Proof: We will show that any transformation that can not be applied to j or k if $t > 1$ can not be applied to j or k if $t = 1$. Transformations which do not involve either j or k which can be applied if $t = 1$ can be applied if $t > 1$, trivially.

Assume there is at least one multi-arc path from j to k and $t > 1$. $Outdeg(j) > 2$ and $indeg(k) > 2$. The transformations which can not be applied are:

(1) an a which requires that *outdeg* $(j) = 1$.

(2) a c which requires that *outdeg* $(j) = 0$,

(3) an a which requires that *indeg* $(k) = 1$,

(4) a c which requires that *indeg* $(k) = 1$.

Assume there is at least one multi-arc path from j to k and $t = 1$. *Outdeg* $(j) > 1$ and *indeg* $(k) > 1$. None of the above transformations could be applied.

Lemma 3.9: The sequence S can be changed to a sequence S' which also collapses G, such that for every $b(j, k)$ in S with t arcs from j to k and one or more paths of length greater than one arc from j to k, $b(j, k)$ is deleted from S'.

Proof: In S, after $b(j, k)$ is applied, there is a single arc (j, k) from j to k, and one or more multi-arc paths from j to k. By lemma 3.7, these paths are all reduced to $q \geq 1$ single arcs from j' to k', where j may have been merged with nodes $j_1 \cdots j_n$ to form j' and k may have been merged with $k_1 \cdots k_m$ to form k'. When $j_1 \cdots j_n$ are merged with j and $k_1 \cdots k_m$ are merged with k, arc (j, k) becomes arc (j', k'). Since there are $q + 1$ or at least two arcs from j' to k' $b(j', k')$ can be applied. $b(j', k')$ replaces the $q + 1$ arcs by a single arc.

We will show that if $b(j, k)$ is not applied, the same result is achieved. In S', if $b(j, k)$ is not applied, there are t arcs from j to k and one or more multi-arc paths from j to k. By Lemma 3.8, any transformation which could be applied with one arc joining j and k can also be applied with t arcs joining j and k. Therefore, the same sequence of transformations which collapse the multi-arc paths from j to k with one arc (j, k) can also be used to collapse the paths with t arcs joining j and k. When nodes $j_1 \cdots j_n$ are merged with j and nodes $k_1 \cdots k_m$ are merged with k, the t arcs from j to k become t arcs from j' to k'. Since there are $q + t$, or at least two arcs from j' to k', $b(j', k')$ can be applied to replace the $q + t$ arcs by a single arc.

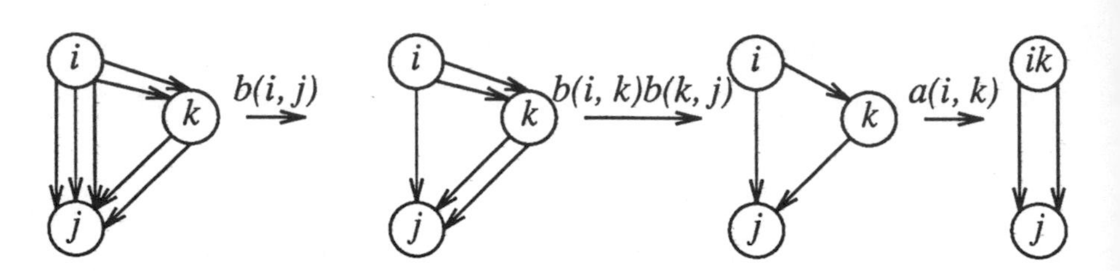

Figure 7-7 $b(i, j)$ applies twice to the same two nodes in sequence S

Lemma 3.10: There is no $b(i, j)$ in S' which is not followed later in S' by either an $a(i, j)$ or $c(i, j)$.

Proof: In S' every arc is deleted. Thus for every arc (i, j) which is introduced by a $b(i, j)$, there must be an occurrence of $a(i, j)$ or $c(i, j)$ later in S' which deletes (i, j).

Lemma 3.11: The sequence S' can be changed to a sequence S'' which also collapses G, and has the following property.

The first a in S'' which applies to either arc (i,˜j) or (j, k), in a subgraph with three nodes i, j, and k such that i is a parent of j, j is a parent of k, $indeg(j) = outdeg(j) = 1$, applies to an arc (either (i, j) or (j, k)) which was introduced by a b, if either arc was so introduced.

Proof: By Lemma 3.4, applying $a(i,j)$ produces the same subgraph as applying $a(j,k)$, so one can be substituted for the other without affecting the rest of the sequence.

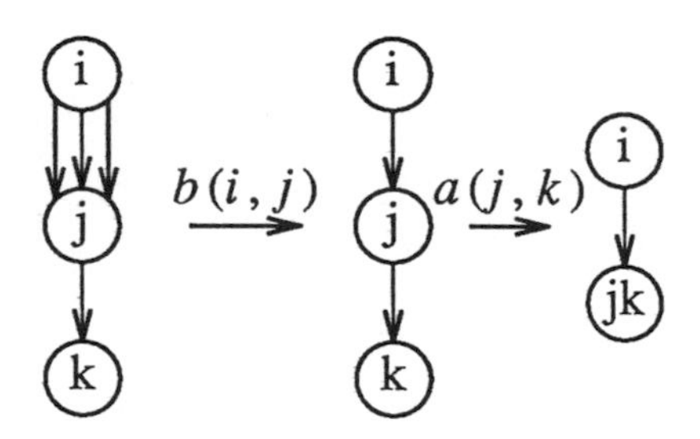

Figure 7-8 Sequence S' in which b (i , j) is followed by a (j , k)

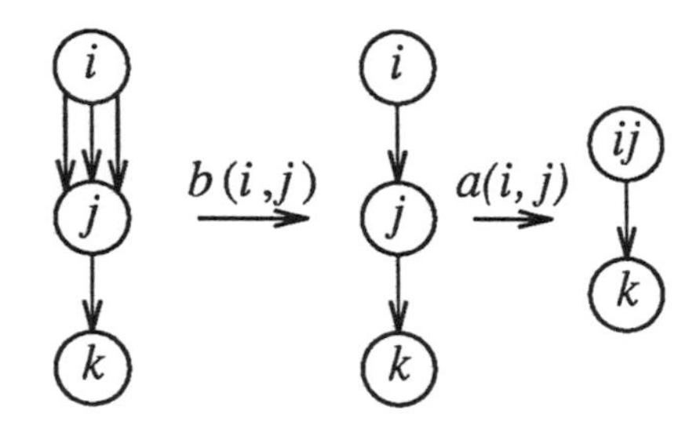

Figure 7-9 Sequence S" in which b (i , j) is followed by a (i , j)

Lemma 3.12: Sequence S'' can be changed to a sequence S''' which also collapses G as follows. In S'', every $b(j, k)$ which introduces arc (j, k) is moved forward in the sequence until it is just before the $a(j, k)$ or $c(j, k)$ which deletes arc (j, k).

Proof: The only effect of $b(j, k)$ is to decrease *outdeg* (j) and *indeg* (k). If the requirements for t, where t is one of $\{a, b, c\}$, do not require either *outdeg* (j) or *indeg* (k) to have been decreased by $b(j, k)$, then $b(j, k)$ can be moved from before t in S'' to after t.

Suppose t does not involve either j or k. Call the two nodes to which t is applied x and y. The requirements for t are only that x and y have a certain indegree and outdegree. Since $b(j, k)$ does not change indegree and outdegree of x and y, $b(j, k)$ can be moved over t.

Suppose t involves either j or k or both j and k.

Case 1: $t = c(x, y)$: In order for $b(j, k)$ to be applied j must be a parent of k and there must be more than one arc from j to k. After $b(j, k)$, there is one arc from j to k. In order for $c(x, y)$ to be applied immediately after $b(j, k)$, x must be a parent of y, $indeg(y) = 1$, $outdeg(y) = 0$.

(1) $y = j$: $c(x, y)$ cannot be applied after $b(j, k)$ since $outdeg(j) = 1$.

(2) $x \neq j$ and $y = k$: $c(x, y) = c(x, k)$ cannot be applied after $b(j, k)$ because x and j are both parents of k, so indeg(k) > 1.

(3) $x = j$ and y $\neq k$: $b(j, k)$ does not change $indeg(y)$ or $outdeg(y)$. Therefore, if $c(x, y) = c(j, y)$ can be applied after $b(j, k)$, it can be applied before $b(j, k)$.

(4) $x = k$: $b(j, k)$ does not change $indeg(y)$ or $outdeg(y)$. Therefore, if $c(x, y) = c(k, y)$ can be applied after $b(j, k)$ it can be applied before $b(j, k)$.

(5) $x = j$, y $= k$: in forming S''', $b(j, k)$ should not be moved over $c(j, k)$.

Case 2: $t = b(x, y)$: The conditions for $b(x, y)$ require the existence of more than one arc from j to k, and conditions for $b(x, y)$ only involve the existence of more than one arc from x to y. Since $b(j, k)$ only affects the arcs from j to k, conditions for b(x,˜y) are not affected by $b(j, k)$ unless $x = j$ and $y = k$. The sequence $b(j, k)b(j, k)$ cannot occur in S'' because, after the first $b(j, k)$, there is only one arc from$j$ to k. Therefore, $b(j, k)$ can be moved over $b(x, y)$.

Case 3: $t = a(x, y)$: requirements for $a(x, y)$ are that x is a parent of y and one of x and y has indeg = 1 and outdeg = 1.

(1) $y = j$: If j has more than one parent or more than one child, *indeg* $(x) \leq 1$ and *outdeg* $(x) = 1$, or $a(x, j)$ could not be applied. $b(j, k)$ does not change *indeg* (x) or *outdeg* (x), so $b(j, k)$ can be moved over $a(x, j)$. If j has indeg = 1 and outdeg = 1, after $b(j, k)$ is applied the subgraph consisting of x, j, k meets the conditions of lemma 3.11, and hence $a(x, j)$ must apply to an arc (x, j) previously introduced by $b(x, j)$. This is impossible if the original graph satisfied restriction 1, since it would require that *indeg* $(j) > 1$ for $b(x, j)$ and *outdeg* $(j) > 1$ for $b(j, k)$.

(2) $x = k$: this case is similar to (1) above.

(3) $x = j$ and $y \neq k$, where y had indeg = 1 and outdeg = 1: $b(j, k)$ does not change indeg or outdeg of y, so $b(j, k)$ can be moved over $a(j, y)$.

(4) $y = k$ and $x \neq j$, where x has indeg = 1 and outdeg = 1: $b(j, k)$ does not change indeg or outdeg of x, so $b(j, k)$ can be moved over $a(x, k)$.

(5) $x = j$ and $y = k$: In forming sequence S''', $b(j, k)$ should not be moved over $a(j, k)$.

Theorem 3.4: Every sequence S''' can be replaced by a sequence V consisting only of A and B which also collapses G.

Proof: By lemma 3.10 there is no $b(i, j)$ in S' to an $a(i, j)$ or $c(i, j)$ later in S'. Therefore, there is no $b(i, j)$ in S''' which is not immediately followed by $a(i, j)$ or $c(i, j)$. By Lemma 3.5 every consecutive pair of transformations $b(i, j)a(i, j)$ can be replaced by $B(i, j)$. By Lemma 3.6 every consecutive pair of transformations $b(i,j)c(i,j)$ can be replaced by $A(i,j)$. The resulting sequence S'''' contains no occurrence of b. By Lemma 3.2 every a in S'''' can be replaced by B. By Lemma 3.3 every c in S'''' can be replaced by A. The resulting sequence V contains only A and B.

Appendix E

TIME BOUNDS OF COLLAPSIBLE GRAPH ALGORITHM

Theorem 3.4: Call the number of nodes in the graph n, the maximum degree of any node in the graph d, and the total number of shapes s. Any collapsible graph G can be collapsed to a single node by transformations A and B in time proportional to $(n-1) \times s^{(d+1)}$.

Proof: All arcs involved in transformation $A(i, j)$ touch node i, both arcs that are eliminated and those which are part of the table for the new node. Each arc can have any of the s shapes assigned to it. The total number of combinations of shapes along arcs that need to be checked, and hence the time for one instance of transformation $A(i, j)$ is s^d where d is the maximum degree node i could have.

All arcs involved in case 1 and case 3 of transformation $B(i, j)$ touch node j except the single arc which enters node i. All arcs involved in case 2 of transformation $B(i, j)$ touch node i except the single arc that leaves node j. Since each arc can be assigned any of the s shapes, the total number of combinations of shapes that need to be checked and hence the total time for an instance of transformation B is $s^{(d+1)}$ where d is the largest possible degree for node j or node i.

Since each A or B collapses 2 nodes into one, G can be collapsed by a sequence of $(n-1)$ A's and B's, each of which takes time $\leq s^{(d+1)}$. So the entire sequence takes time $\leq (n-1) \times s^{(d+1)}$. Also, since neither transformation A or B ever creates a node with larger degree than before, the value of d, the degree of the node with largest degree, is known from the original graph.

Appendix F

COST FUNCTION FOR SHARED NODES IN THE SET COVER PROOFS

The cost of each shared node will be taken to be the number of distinct shapes assigned to arcs which touch the node, minus 1. This function was chosen to be easily computed from a configuration, and to be specifiable without requiring a table whose size grows exponentially with the number of arcs touching a node. The subroutine below demonstrates how such a function could be computed.

Let NSHAPES be a constant which represents the number of shapes. Let the sequence of numbers $\{1 \cdots \text{NSHAPES}\}$ represent the set of shapes. Let shapes_present[NSHAPES] be an array of zeroes and ones. Let i be a node with n incoming arcs, each of which is assigned a shape. These arcs will be labeled $\{i.1 \cdots i.n\}$. The shape assigned to arc $i.j$ will be called $s_{i.j}$. The variable $s_{i.j}$ represents a number between 1 and NSHAPES, indicating which one of $\{1 \cdots$ NSHAPES $\}$ is assigned to arc $i.j$.

```
for k = 1 to NSHAPES
    shapes_present[k] = 0;

for j = 1 to n
    shapes_present[s_{i.j}] = 1;

shapecount = 0;
for k = 1 to NSHAPES
    if shapes_present[k] = 1
    then shapecount = shapecount + 1;

return(shapecount - 1);
```

References

1. A. V. Aho and S. C. Johnson, ''Optimal Code Generation for Expression Trees,'' *JACM* **23**(3), pp. 488-501 (July, 1976).
2. A. V. Aho, S. C. Johnson, and J. D. Ullman, ''Code Generation for Machines with Multiregister Operations,'' *SIGPLAN proceedings*, pp. 21-28 (Jan, 1977).
3. A. V. Aho and J. D. Ullman, *Principles of Compiler Design,* Addison-Wesley (1979).
4. A. V. Aho, J. E. Hopcroft, and J. D. Ullman, *The Design and Analysis of Computer Algorithms,* Addison-Wesley (1976).
5. A. V. Aho, S. C. Johnson, and J. D. Ullman, ''Code Generation for Expressions with Common Subexpressions,'' *JACM* **24**(1), pp. 146-160 (Jan, 1977).

6. Umberto Bertele and Francesco Brioschi, *Nonserial Dynamic Programming,* Academic Press (1972).

7. J. L. Bruno and T. Lassagne, "The Generation of Optimal Code for Stack Machines," *JACM* **22**(3), pp. 2382-396 (July, 1975).

8. Paul Budnick and David J. Kuck, "The Organization and Use of Parallel Memories," *IEEE Transactions on Computers* (December, 1971).

9. S. E. Dreyfus and A. M. Law, *The Art and Theory of Dynamic Programming,* Academic Press (1977).

10. M. R. Garey and D. S. Johnson, *Computers and Intractibility,* W. H. Freeman (1979).

11. M. R. Garey and D. S. Johnson, "Some Simplified NP-Complete Problems," *SIGACT proceedings* (May, 1974).

12. F. Harary, J. Krarup, and Shwenk, A., "Graphs Suppressible to an Edge," *Canadian Math Bulletin* **15**(2), pp. 201-204 (1972).

13. John E. Hopcroft and Jeffrey D. Ullman, *Introduction to Automata Theory, Languages, and Computation,* Addison-Wesley (1979).

14. Richard M. Karp, "Reducibility among Combinatorial Problems," University of California Tech Report 3 (April, 1972).

15. M. J. Kascic, "Vector Processing on the Cyber 200," *Infotech state of the art report: "Supercomputers"* (1979).

16. David J. Kuck, *The structure of Computers and Computations,* John Wiley and Sons (1978).

17. Eugene L. Lawler, Robert E. Tarjan, and Jacobo Valdes,

"The Recognition of Series Parallel Digraphs," *11'th Annual Symposium on Theory of Computing* (May, 1979).

18. Duncan H. Lawrie, "Access and Alignment of Data in an Array Processor," *IEEE Transactions on Computers* **c-24**(12) (December, 1975).

19. Michael Leuze, *Memory Access Patterns in Vector Computers with Applications to Problems in Linear Algebra,* Ph. D. Dissertation, Duke University (December, 1981).

20. G. F. Pfister, W. C. Brantley, D. A. George, S. L. Harvey, W. J. Kleinfelder, K. P. McAuliffe, E. A. Melton, V. A. Norton, and J. Weiss, "The IBM Research Parallel Processor Prototype (RP3)," *Proceedings of the 1985 International Conference on Parallel Processing*, pp. 764-771.

21. Bhaskaram Prabhala and Ravi Sethi, "Efficient Computation of Expressions with Common Subexpressions," *JACM* **27**(1), pp. 146-163 (Jan, 1980).

22. D. S. Rubin, *Introduction to Dynamic Programming*, 1981.

Index

J

L

M

N

O

P

R

S

T

V